Advanced Piping Design

Advanced Piping Design

Editor

Rishi Raj

Advanced Piping Design

Edited by **Rishi Raj**

Printed in 2017

ISBN: 978-1-68117-329-0

Library of Congress Control Number: 2015939241

© 2016 by
SCITUS Academics LLC,
616, Corporate Way, Suite 2, 4766,
Valley Cottage, NY 10989

www.scitusacademics.com

Contents

Preface

The Fundamentals of Piping Design, the objective was to arm the reader with the basic "rules" for the design, fabrication, installation, and testing of process and utility piping systems for oil and gas refineries, chemical complexes, and production facilities at both offshore and onshore locations. The objective of Volume 2, Advanced Piping Design, on the same subject, is to look into more detail at the design of process piping systems in specific locations around the various items of process equipment that would be typically found in a petrochemical or oil and gas processing facility.We enlisted the direction and support of Rutger Bott,ermans, from Delft in The Netherlands, who is the author of this title. He wrote the text in a very direct style to avoid any misinterpretation. The bullet point/checklist format allows the reader so see quickly if he or she has considered the point when laying out the piping system. Rutger and his company Red-Bag, have a great deal to offer the industry; and I look forward to working with him again on other projects.

Editor

CFD Modeling of Gas-Liquid Bubbly Flow in Horizontal Pipes: Influence of Bubble Coalescence and Breakup

K. Ekambara, R. Sean Sanders, K. Nandakumar, and J. H. Masliyah

Department of Chemical and Materials Engineering, University of Alberta, Edmonton, AB, Canada T6G 2G6

ABSTRACT

Modelling of gas-liquid bubbly flows is achieved by coupling a population balance equation with the three-dimensional, two-fluid, hydrodynamic model. For gas-liquid bubbly flows, an average bubble number density transport equation has been incorporated in the CFD code CFX 5.7 to describe the temporal and spatial evolution of the gas bubbles population. The coalescence and breakage effects of the gas bubbles are modeled. The coalescence by the random collision driven by turbulence and wake entrainment is considered, while for bubble breakage, the impact of turbulent eddies is considered. Local spatial variations of the gas volume fraction, interfacial area concentration, Sauter mean bubble diameter, and liquid velocity are compared against experimental data in a horizontal pipe, covering a range of gas (0.25 to 1.34 m/s) and liquid (3.74 to 5.1 m/s) superficial velocities and average volume fractions (4% to 21%). The predicted local variations are in good agreement with the experimental measurements reported in the literature. Furthermore, the development of the flow pattern was examined at three different axial locations of L/D = 25, 148, and 253. The first location is close to the entrance region where the flow is still developing, while the second and the third represent nearly fully developed bubbly flow patterns.

INTRODUCTION

Gas-liquid, two-phase flow in horizontal pipes is encountered often in a number of industrial processes. Common applications include chemical plants, evaporators, oil wells and pipelines, fluidized bed combustors, and evaporators. Horizontal bubbly flows have received less attention in the literature than vertical flows, even though this flow orientation is equally important in industrial applications such as hydrotransport, an important technology in bitumen extraction. Experimental observations are also difficult in this case, as the migration of dispersed bubbles towards the top of the pipe, due to buoyancy, causes a highly nonsymmetric volume distribution in the pipe cross-section. This density stratification is not often accompanied by a strong secondary flow. Gas volume fraction, interfacial area concentration, and mean bubble diameter are the three characterizing field variables that characterize the internal flow structure of two-phase, gas-liquid flows in horizontal pipe. In various industrial processes, the gas volume fraction parameter is required for hydrodynamic and thermal design. The interfacial transport of mass, momentum, and energy is proportional to the interfacial area and the driving forces. This is an important parameter required for a two-fluid model formulation. The mean bubble diameter serves as a link between the gas volume fraction and interfacial area concentration. An accurate knowledge of local distributions of these three parameters is of great importance to the eventual understanding and modelling of the interfacial transfer processes [1].

In the past two decades, significant developments in the modeling of two-phase flow processes have occurred since the introduction of the two-fluid model. In the volume averaged, two-fluid model, the interfacial transfer terms are strongly related to the interfacial area concentration and the local transfer mechanisms such as the degree of turbulence near the interfaces. Fundamentally, the interfacial transport of mass, momentum, and energy are proportional to the interfacial area concentration (a_{ij}) and driving forces. Since the interfacial area concentration a_{ij} represents the key parameter that links the interaction of the phases, significant attention has been paid towards developing a better understanding of the coalescence and breakage effects due to interactions among bubbles and between bubbles and turbulent eddies for gas-liquid bubbly flows [2–5]. The population balance method is a well-known method for tracking the size distribution of the dispersed phase and accounting for the breakage and coalescence effects in bubbly flows (see, e.g. [6–14]). This approach is concerned with maintaining a record of the number of bubbles initially and tracking their evolution in space over time.

In this work, an attempt has been made to demonstrate the possibility of combining population balance with computational fluid dynamics (CFD) for the case of gas-liquid bubbly flow in the horizontal pipe. The MUSIG model has been implemented in CFX-5.7 to account for the nonuniform bubble size distribution in a gas-liquid flow [7, 15, 16]. The gas volume fraction, interfacial area concentration, Suater mean diameter, and axial liquid velocity have been predicted for a wide range of gas and liquid flow condition. Further, the development of flow pattern has been studied at three different axial locations. The model predictions are compared with available experimental data from the literature.

MATHEMATICAL MODELLING

Population Balance Model

Population balance modelling is used in computing the size distribution of the dispersed phase and in accounting for the breakage and coalescence effects in bubbly flows. A general form of the population balance equation is

$$\frac{\partial n_i}{\partial t} + \nabla \cdot \left(\mathbf{u}_g n_i \right) = B_B + B_C - D_B - D_C,$$

(1)

where u_g is the gas velocity, n_i represents the number density of size group i, and terms on the right hand side B_B, B_C, D_B, and D_C are, respectively, the *"birth"* and *"death"* due to breakup and coalescence of bubbles. The left hand side tracks the spatial and temporal evolution of a class of bubbles, while the right hand side models the exchange between classes due to breakup and coalescence of bubbles. The bubble number density n_i is related to the gas volume fraction α_g by

$$\alpha_g f_i = n_i V_i,$$

(2)

where f_i represents the volume fraction of bubbles of group i, and V_i is the corresponding volume of a bubble of group i. It is necessary to provide individual models for each of the breakup and coalescence processes as it depends on the mechanisms and is sensitively dependent on the presence of surfactants, turbulence levels, and so forth. These models are discussed next.

Bubble Breakup Model

The breakup of bubbles in turbulent dispersions employs the model developed by Luo and Svendsen [18]. Binary break-up of the bubbles is assumed, and the model is based on the theories of isotropic turbulence. For binary breakage, a dimensionless variable describing the sizes of daughter drops or bubbles (the breakage volume fraction) can be defined as

$$f_{BV} = \frac{V_i}{V} = \frac{d_i^3}{d^3} = \frac{d_i^3}{d_i^3 + d_j^3},$$

(3)

where d_i and d_j are diameters (corresponding to V_i and V_j) of the daughter bubbles in the binary breakage of a parent bubble with diameter d (corresponding to volume V). The value interval of the breakage volume fraction is between 0 and 1. The break-up rate of bubbles of volume V_j into volume sizes of $V_i (= V_{fBV})$ can be obtained as

$$
\frac{\Omega\left(V_j : V_i\right)}{\left(1 - \alpha_g\right)n_j}
$$

$$
= C\left(\frac{\in}{d_j^2}\right)^{1/3} \int_{\zeta_{min}}^{1} \frac{(1+\zeta)^2}{\zeta^{11/3}} \exp\left(-\frac{12 c_f \sigma}{\beta \rho_l \in^{2/3} d_j^{5/3} \zeta^{11/3}}\right) d\zeta,
$$

$$(4)$$

where $\in$ is the rate of energy dissipation per unit of liquid mass; $\zeta = \lambda/d_j$ is the size ratio between an eddy and a particle in the inertial subrange and consequently $\zeta_{min} = \lambda_{min}/d_j$; C and β are determined, respectively, from fundamental consideration of drops or bubbles breakage in turbulent dispersion systems to be 0.923 and 2.0 in Luo and Svendsen [18]; c_f is the increase coefficient of surface area given by

$$
c_f = \left[f_{BV}^{2/3} + (1 - f_{BV})^{2/3} - 1 \right].
$$

$$(5)$$

The birth rate of group i bubbles due to break-up of larger bubbles is

$$
B_B = \sum_{j=i+1}^{N} \Omega\left(V_j : V_i\right) n_j.
$$

$$(6)$$

The death rate of group i bubbles due to break-up into smaller bubbles is

$$
D_B = \Omega_i n_i \text{ with } \Omega_i = \sum_{k=1}^{i} \Omega_{ki}.
$$

$$(7)$$

Bubble Coalescence Model

The coalescence of two bubbles is assumed to occur in three steps. The first step where the bubbles collide and trap a layer of liquid between them, a second step where this liquid layer drains until it reaches a critical thickness, and a last step during which this liquid film disappears and the bubbles coalesce. The collisions between bubbles may be caused by turbulence, buoyancy, or laminar shear. Only the first cause of collision (turbulence) is considered in the present model. Indeed collisions caused by buoyancy cannot be taken into account as all the bubbles from each class move at the same speed. The coalescence rate considering turbulent collision taken from Prince and Blanch [19] can be expressed as

$$\chi = \theta_{ij} \exp\left(-\frac{t_{ij}}{\tau_{ij}}\right),$$

(8)

where t_{ij} is the contact time for two bubbles given by $(dij/2)^{2/3}/\epsilon^{1/3}$.

When bubbles collide, a small amount of liquid is entrapped between them, forming a small circular lens or film of radius R and thickness h. The forces causing the film or lens to grow thinner in pure systems arise from capillary pressure, augmented by compression from a close range Hamaker force which accounts for the mutual attraction of water molecules on opposite sides of the liquid film [19]. For equal size bubbles, Oolman and Blanch [20] derived the thinning formula

$$\frac{-dh}{dt} = \left\{\frac{8}{R^2\rho_L}\left[h^2\left(\frac{2\sigma}{r_b} + \frac{A}{6\pi h^3}\right)\right]\right\}^{1/2}.$$

(9)

Prince and Blanch [19] solved the above equation numerically and show that t_{ij}, the time required for two bubbles, having diameters di and d_j to coalesce is estimated to be $\{(d_{ij}/2)^3\rho l/16\sigma\}^{1/2} \ln(h_0/h_j)$. The equivalent diameter d_{ij} is calculated as suggested by Chesters and Hoffman [21]: $d_{ij} = (2/d_i + 2/d_j)^{-1}$. The parameters h_0 and h_j represent the film thickness when collision begins and critical film thickness at which rupture occurs, respectively. The values of the above parameters depend mainly on the physical properties of the liquid phase and have been experimentally computed for the air-water system. According to Prince and Blanch [19], for air-water systems, experiments have determined h_0 and h_j to be 1×10^{-4} m [22] and 1×10^{-8} m [23], respectively.

The turbulent collision rate θ_{ij} for two bubbles of diameters d_i and d_j is given by

$$\theta_{ij} = \frac{\pi}{4} \left[d_i + d_j \right]^2 \left(\mathbf{u}_{ti}^2 + \mathbf{u}_{tj}^2 \right)^{1/2},$$

(10)

where the turbulent velocity $\mathbf{u}_t$ in the inertial subrange of isotropic turbulence [24] is,

$$\mathbf{u}_t = 1.4 \in^{1/3} d^{1/3}.$$

(11)

The birth rate of group i due to coalescence of group k and group l bubbles is:

$$B_C = \frac{1}{2} \sum_{k=1}^{N} \sum_{l=1}^{N} \chi_{i,kl} n_i n_j.$$

(12)

The death rate of group i due to coalescence with other bubbles is:

$$D_C = \sum_{j=1}^{N} \chi_{ij} n_i n_j.$$

(13)

Flow Equations

The numerical simulations presented are based on the two-fluid, Eulerian-Eulerian model. The Eulerian modelling framework is based on ensemble-averaged mass and momentum transport equations for each phase. Regarding the liquid phase (α_l) as the continuum and the gaseous phase (bubbles) as the dispersed phase (α_g), these equations without interface mass transfer can be written in standard form as follows.

Continuity equation of the liquid phase

$$\frac{\partial}{\partial t}(\rho_l \alpha_l) + \nabla \cdot (\rho_l \alpha_l \mathbf{u}_l) = 0.$$

(14)

Continuity equation of the gas phase

$$\frac{\partial}{\partial t}\left(\rho_g \alpha_g f_i\right) + \nabla \cdot \left(\rho_g \alpha_g \mathbf{u}_g f_i\right) = S_i.$$

(15)

Momentum equation

$$\frac{\partial}{\partial t}(\rho_k \alpha_k \mathbf{u}_k) + \nabla \cdot (\rho_k \alpha_k \mathbf{u}_k \mathbf{u}_k) = -\alpha_k \nabla p + \rho_k \alpha_k g_i + \alpha_k \mu_k \nabla^2 \mathbf{u}$$

$$+ \mathbf{F}_{km} \quad (k, \; m = l, \; g).$$

(16)

In (15), S_i is a source term that captures the coalescence and break-up processes. The right side of (16) describes the following forces acting on the phase k: the pressure gradient, gravity, and the viscous stress term, and $\mathbf{F}_{km}$ represents the sum of the interfacial forces that include the drag force $\mathbf{F}_D$, the lift force $\mathbf{F}_L$, the virtual mass force $\mathbf{F}_{VM}$, the wall lubrication force $\mathbf{F}_{WL}$, and the turbulent dispersion force $\mathbf{F}_{TD}$. Detailed descriptions of each of these forces can be found in Anglart and Nylund [25]; Lahey and Drew [26], and Joshi [27].

The origin of the drag force is due to the resistance experienced by a body moving in the liquid. Viscous stress creates skin drag, and pressure distribution around the moving body creates form drag. The formulation of the drag force is a key issue in multiphase flows. Clift et al. [28] and Joshi et al. [29] have given excellent accounts of this subject. The interphase momentum transfer between gas and liquid due to drag force is given by

$$\mathbf{F}_D = \frac{3}{4} C_D \alpha_g \rho_l \frac{1}{d_S} |\mathbf{u}_l - \mathbf{u}_g| (\mathbf{u}_l - \mathbf{u}_g).$$

(17)

The lift force considers the interaction of the bubble with the shear field of the liquid. It acts perpendicular to the main flow direction and is proportional to the gradient of the liquid velocity field. The lift force in terms of the slip velocity and the curl of the liquid phase velocity can be modelled as [30–33]

$$\mathbf{F}_L = C_L \alpha_g \rho_l \left(\mathbf{u}_g - \mathbf{u}_l \right) \times \nabla \times \mathbf{u}_l.$$

(18)

The wall lubrication force arises because the liquid flow rate between bubble and the wall is lower than between the bubble and the main flow. This results in a hydrodynamics pressure difference driving bubble away from the wall. This force density is approximated as [34]

$$\mathbf{F}_{WL} = -\alpha_g \rho_l \frac{(\mathbf{u}_r - (\mathbf{u}_r \cdot n_w)n_w)}{d_S} \max \left[C_1 + C_2 \frac{d_S}{y_w}, 0 \right].$$

(19)

Here, $\mathbf{u}_r = \mathbf{u}_l - \mathbf{u}_g$ is the relative velocity between phases, d_s is the dispersed phase Sauter mean bubble diameter, y_w is the distance to the nearest wall, and n_w is the unit normal pointing away from the wall.

The turbulent dispersion force, derived by Lopez de Bertodano [35], is based on an analogy with molecular movement. The turbulence-induced dispersion is a function of turbulent kinetic energy and gradient of the volume fraction of the liquid:

$$\mathbf{F}_{TD} = -C_{TD} \rho_l k \nabla \alpha_l.$$

(20)

The drag coefficient C_D in (17) has been modelled using Ishii and Zuber [36] drag model. The lift coefficient C_L is theoretically proven to be 0.5 for a spherical bubble in a potential flow [37]. It is also known that (i) C_L becomes smaller than 0.5 for a single small bubble in a viscous flow, as show by Lopez de Bertodano et al. [38] and Lance and Lopez de Bertodano, [39] (C_L = 0.25; 0.1, resp.), and (ii) C_L strongly depends on the bubble diameter and decreases with d_s [40]. These facts indicate that C_L is a function of bubble diameter and fluid properties. From this point of view, it is necessary to consider the lift coefficient also to depend on flow conditions. Recently, Tomiyama et al. [41] have developed an empirical correlation for the lift coefficient as a function of Reynolds number and Eotvos number. We have found that this correlation does not perform well for horizontal flows because of the migration of dispersed bubbles towards the top of the pipe. In view of this, we have developed a correlation for lift coefficient in horizontal flows. An interesting finding and a main contribution in this work is that a wide range of flow behavior of two-phase bubbly flows in horizontal pipes is represented by a unique functional relationship between the lift coefficient and the flow Reynolds number. When such closure relations are tested over a wide range of two-phase flows (not only pipe flows, but also bubble columns, etc.) our confidence in using such models to study practical two phase problems in process equipment will increase over a period of time. Further explanation about C_L is given in the results and discussion. The wall lubrication constants C_1 and C_2, as suggested by Antal et al., [34], are −0.01 and 0.05, respectively. The coefficient $C_{TD} = 0.5$ was found to give the good results which is in the recommended range of 0.1 to 1.0 [35]. By definition, the interfacial area concentration a_{ij} for bubbly flows can be determined through the relationship

$$a_{ij} = \frac{6\alpha_g}{d_S},$$

(21)

where d_S is the bubble Sauter mean diameter. The local bubble Sauter mean diameter based on the calculated values of the scalar fraction f_i and discrete bubble sizes d_i can be deduced from

$$d_S = \frac{1}{\sum_i f_i/d_i}.$$

(22)

From the drag and nondrag forces above, it is evident that the interfacial area concentration a_{ij} and the bubble Sauter mean bubble diameter in (22) are essential parameters that link the interaction between the liquid and gas (bubbly) phases. In most two-phase flow studies, the common approach of prescribing constant bubble sizes through the mean bubble Sauter diameter is still prevalent. Such an approach does not allow dynamic representation of the changes in the interfacial structure.

Turbulence Equations

For the continuous liquid phase, a k-$\in$ model is applied with its standard constants: $C_{\in 1}$ = 1.44, $C_{\in 2}$ = 1.92, C_{μ} = 0.09, σ_k = 1.0, and $_{\in}$ = 1.3. No turbulence model is applied on the dispersed gas phase, but the influence of the dispersed phase on the turbulence of the continuous phase is taken into account with Sato's additional term [42]. The governing equations for the turbulent kinetic energy k and turbulent dissipation ε are

$$\frac{\partial}{\partial t}(\rho_l \alpha_l k) + \frac{\partial}{\partial x_i}(\rho_l \alpha_l \mathbf{u}_l k) = \frac{\partial}{\partial x_i}\left(\alpha_l\left(\mu_l + \frac{\mu_{l,\text{tur}}}{\sigma_k}\right)\frac{\partial k}{\partial x_j}\right)$$
$$+ \alpha_l(G - \alpha_l \rho_l \in),$$

$$\frac{\partial}{\partial t}(\rho_l \alpha_l \in_l) + \frac{\partial}{\partial x_i}(\rho_l \alpha_l \mathbf{u}_l \in) = \frac{\partial}{\partial x_i}\left(\alpha_l\left(\mu_l + \frac{\mu_{l,\text{tur}}}{\sigma_\in}\right)\frac{\partial \in}{\partial x_i}\right)$$
$$+ \alpha_l \frac{\in}{k}(C_{\varepsilon 1}G - C_{\varepsilon 2}\alpha_l \rho_l \in), \qquad (23)$$

where G is the turbulence production due to viscous and buoyancy forces, which is modeled using

$$G = \mu_t \nabla \mathbf{u} \cdot \left(\nabla \mathbf{u} + \nabla \mathbf{u}^T\right) - \frac{2}{3}\nabla \cdot \mathbf{u}(3\mu_t \nabla \cdot \mathbf{u} + \rho_l k)$$
$$- \frac{\mu_t}{\rho \sigma_p}\rho_l \beta g \cdot \nabla T. \qquad (24)$$

METHOD OF SOLUTION

The multiple size group (MUSIG) model (CFX 5.7 from ANSYS) used in this study combines the population balance method with the break-up [18] and coalescence [19] models in order to predict the bubble size distribution of the dispersed phase, and it uses the Eulerian-Eulerian two-fluid model. A standard two-phase flow calculation, with equation for continuity, momentum, and turbulence for a continuous and a dispersed phase, can be extended to include mass fraction of bubbles within several size ranges using the MUSIG model. The size range of the bubbles is split into several groups with, for example, bands of equal diameter or equal volume. Equations are solved for the mass fraction in each band. The MUSIG model has been implemented in the CFX-5.7 software to account nonuniform bubble size distribution in a gas-liquid mixture. The MUSIG model has been extensively used for different systems [7, 10, 15, 16, 43, 44]. These size fractions provide a more accurate measure of the interfacial area density and therefore allow better calculation of the heat and mass transfer taking place between the continuous and dispersed phases.

In this present study, bubbles ranging from 1 mm to 10 mm diameter are equally divided into 10 classes (see Table 1) as the experimental observation of maximum bubble diameter for highest superficial gas velocity is 6 mm. Even if we have considered the range, model predictions picks up the experimental observation bubble size range. The fate of the discrete bubble sizes so prescribed was tracked using the population balance model. Instead of considering 11 different complete phases, it was assumed that each bubble class travels at the same mean algebraic velocity to reduce the computational time and cost. Therefore, it results in 10 continuity equations for the gas phase coupled with a single continuity equation for the liquid phase. Sensitivity of the number of size groups needed to describe a meaningful distribution was examined by dividing the bubble diameters equally into 10, 15, and 20 size groups. The results revealed that no appreciable difference ($\pm$ 2 %) was found for the predicted maximum bubble Sauter mean diameter using the 10, 15, or 20 bubble size groups. For the subdivision into 10 size groups, the maximum Sauter bubble diameter was under predicted by a maximum difference of 2%. In view of computational resources and times, it was therefore concluded that the subdivision of the bubbles sizes into 10 size groups was sufficient and all subsequent computational results are based on the discretization of the bubble sizes into 10 groups.

Table 1: Diameter of each bubble class tracked in the simulation

Class index	1	2	3	4	5	6	7	8	9	10
Bubble diameter, d_i(mm)	1.45	2.35	3.25	4.15	5.05	5.95	6.85	7.75	8.65	9.55

Solution to the coupled sets of governing equations for the balances of mass and momentum of each phase was obtained using CFX 5.7. The conservation equations were discretized using the control volume technique. Computational grid is based on the unstructured set of blocks each containing structured grid. The structured grid within each block is generated using general curvilinear coordinates ensuring accurate representation of the flow boundaries. In order to select an adequate grid resolution, the effect of changing grid size was investigated. Several simulations were carried out using progressively larger number of grid points of 87156, 152482, 257670, 303245, and 341612. Sample grid sensitivity results are shown in Figure 1. It can be seen that there is practically no change in the gas volume fraction and liquid velocity profiles when the grid size increased beyond 257670. In view of the observed effect of grid size, the simulations have been carried out by using 257670 grid points. Initial simulations were carried out with a coarse mesh to obtain an initially converged solution and to obtain an indication of where a high mesh density was needed. However, a dense mesh required additional computational effort. The velocity-pressure linkage was handled through the SIMPLE procedure. Three-dimensional transient simulations were performed. The time stepping strategy used in the transient simulations to reach a steady state was typically a variable step size strategy according to the following scheme: 100 steps at 1×10^{-4} s, followed by 300 steps at 5×10^{-4} s, 400 steps at 1×10^{-3} s, 1400 steps at 5×10^{-3} s, and 8000 steps at 1×10^{-2} s. Underrelaxation factors between 0.6 and 0.7 were adopted for all flow quantities, and pressure was never underrelaxed. The hybrid-upwind discretization scheme was used for the convective terms. At the inlet, gas, liquid, and the average volume fraction have been specified. At the pipe outlet, a relative average static pressure of zero was specified. For initiating the numerical solution, average volume fraction and parabolic liquid velocity profile are specified as initial conditions. The operating conditions are summarized in Table 2. The liquid and gas superficial velocities were varied between 3.74 to 5.1 m/s and 0.25 to 1.34 m/s, respectively. The fluid data are taken at room temperature (25°C) and are treated as isothermal and saturated. Therefore, heat and mass transfer effects are neglected in the simulations.

Table 2: Operating conditions

Geometry	50.3 mm ID
Gas phase	Air at 25°C
Liquid phase	Water at 25°C
Gas superficial velocity	0.25–1.34 m/s
Liquid superficial velocity	3.74–5.1 m/s
Average gas volume fraction	0.04–0.205

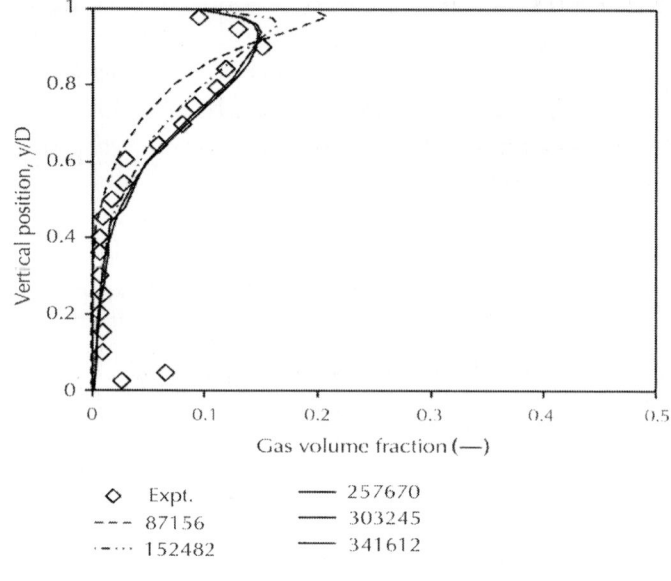

(a)

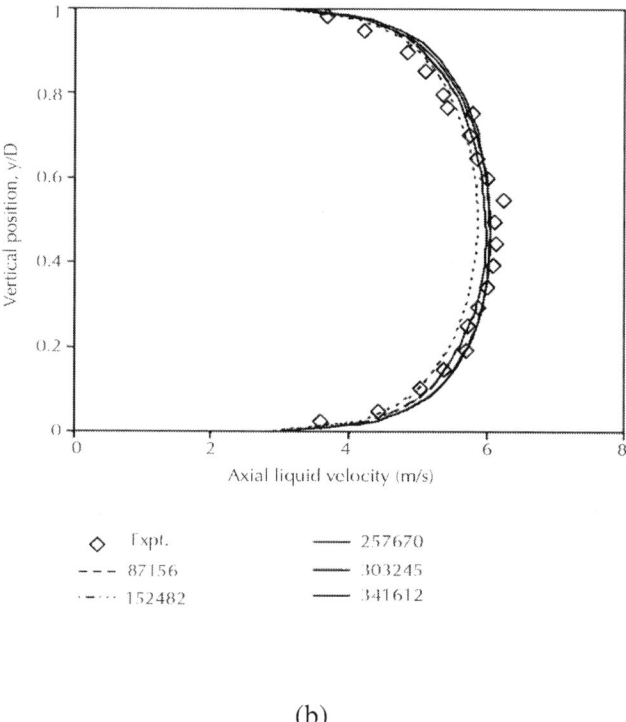

(b)

Figure 1: Effect of grid size on gas volume fraction and axial liquid velocity.

RESULTS AND DISCUSSION

The CFD simulations are carried out for the experimental conditions reported by Kocamustafaogullari and Wang [17] and Kocamustafaogullari and Huang [1]. The local radial profiles of the gas volume fraction, interfacial area concentration, Sauter mean bubble diameter, and liquid velocities are predicted by solving the coupled two-fluid and population balance models. The predicted results are compared with the experimental data at the axial location of L/D = 253 and along a vertical and horizontal line passing through the centre of the pipe axis. Here, y/D and x/R are the normalized vertical and horizontal positions in the pipe.

Estimation of Lift Coefficient

Accurate prediction of developing bubbly flows in horizontal pipes cannot be carried out without sufficient knowledge of a transverse lift force acting on the bubbles, the force that governs the transverse migration of a bubble in a shear field. It has been clarified through a number of experiments that the lateral migration strongly depends on bubble size, that is, small bubbles tend to migrate toward the pipe wall which results in a peak in the bubble volume fraction distribution near the wall. Just like the functional form of the drag coefficient for a single particle interaction is extended to multiparticle systems, the functional form of the lift force that captures the lateral migration phenomenon is given by (18) which has C_L as the lift coefficient. Just as the drag coefficient is a function of local Reynolds number based on the slip velocity, one can expect the lift coefficient also to vary with local Reynolds number, and in general, it is an unknown function for such a complex flow field. In the literature, it is used as a fitting parameter, but various values have been reported. Further, the most of the correlations available in the literature were for vertical flows. The correlations of Legendre and Magnaudet [45], Tomiyama et al. [41] have been used in the simulations and the simulation results shown in Figure 2(a). It can be observed that the correlation of Legendre and Magnaudet [45], Tomiyama et al. [41] does not perform well for the horizontal flows because of most the dispersed bubbles migrate towards the top of the pipe, due to buoyancy. The negative lift coefficient needed because this force pushes bubbles to the pipe center. For the given simulation, we need a negative lift coefficient to predict near wall peak for the gas volume fraction profile. We need a correlation which gives negative lift coefficient value. In view of this, we have developed a correlation to be a function of Reynolds number.

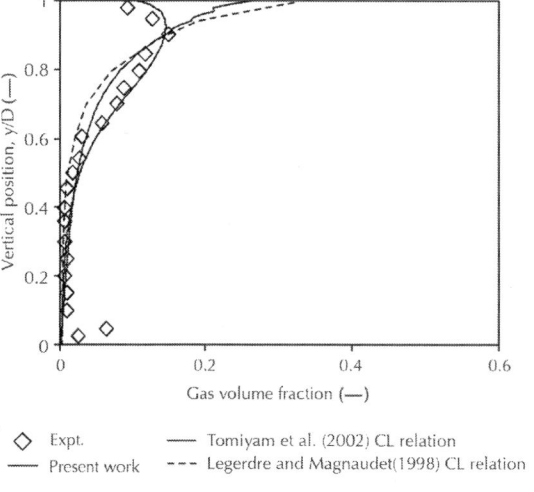

(a)

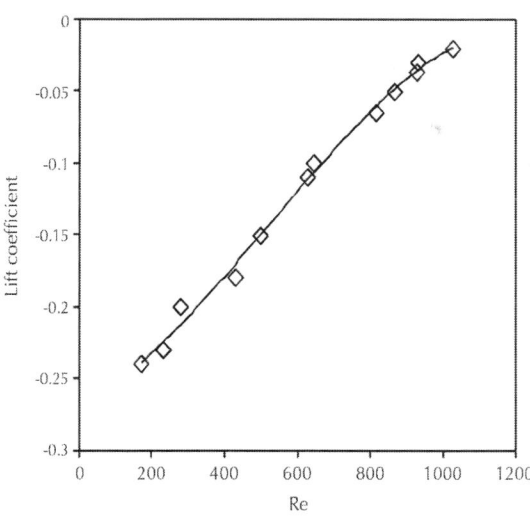

(b)

Figure 2: Lift coefficient (C_L).

The difference between the model predictions and experimental data on the spatial variation of field quantities such as liquid velocity profiles, volume fraction profiles, and interfacial area measurements is minimized by the tuning of this parameter. Bubbly horizontal pipe flow experiments by Kocamustafaogullari and Wang [17]; Kocamustafaogullari and Huang [1]; Kocamustafaogullari et al. [46] were chosen for tuning the lift coefficient as they have detailed experimental data on the spatial variation of liquid velocity profiles, volume fraction profiles, and interfacial area measurements. In all the experiments, adiabatic, incompressible, air-water bubbly flows at atmospheric pressure and room temperature were used. The main result of tuning this parameter is shown in Figure 2(b). The estimated values of the lift coefficient at different experimental condition of gas and liquid flow rates were not scattered all over, but exhibited a well-defined correlation with the Reynolds number defined as $Re = d_s V_s \rho_L / \mu_L$, where d_s is the average bubble diameter, and V_s is the slip velocity. We capture this relationship by a polynomial expression of the form, $C_L = a\,Re^3 + b\,Re^2 + c\,Re + d$, where $a = -1 \times 10^{-10}$, $b = 2 \times 10^{-7}$, $c = 2 \times 10^{-4}$, and $d = -0.2937$. It is worth noting that the correlation is based on the locally measured properties of turbulence as well as the bubble number density, and hence, one can expect it to be valid irrespective of the dimension of the pipe as well as the liquid system. Such a relationship can then be used back in the simulation for predictive purposes at other flow conditions

Gas Volume Fraction

Figures 3(a)–6(a) show the comparison of the predicted gas volume fraction with the experimental data of Kocamustafaogullari and Wang [17] for different superficial gas velocities of 0.25 m/s, 0.50 m/s, 0.80 m/s, and 1.34 m/s at a fixed liquid velocity of 5.1 m/s. Similarly Figures 7(a) and 8(a) show the gas volume fraction for liquid velocities of 3.74 m/s and 4.40 m/s at a fixed gas velocity of 0.51 m/s. The agreement between the predicted and the experimental profiles can be seen to be very good. As the superficial gas velocity increases, the average gas volume fraction also increases. It can be observed from these figures that most of the bubbles tend to migrate towards the top of the pipe wall under the dominating influence of buoyancy force. The balance of buoyancy and lift forces causes the profiles of gas volume fraction to show a distinct peak near the top wall at about y/D = 0.9 to 0.95 for all the flow conditions. A similar observation was made experimentally by Kocamustafaogullari and Wang [17], Kocamustafaogullari and Huang [1], Kocamustafaogullari et al., [46], and Iskandrani and Kojasoy [47]. At a constant gas superficial velocity of 0.51 m/s in Figures 7 and 8, the average and the peak value of the volume fraction decreases with increasing liquid velocity, as expected. The fact that the spatial variation of the gas volume fraction matches well with the experimental data over a wide range of flow conditions gives us confidence that the lift coefficient correlation that we have developed is quite appropriate. The real test of this correlation must of course await testing against similar data in a larger diameter pipe. The challenge in developing multiphase flow models using the volume-averaged framework is to develop such closure relationships for each of the individual mechanisms and test their validity under a wide variety of scales and flow conditions. We will be testing this correlation for bubble columns in the near future. The model prediction of gas volume fraction shows relative mean and maximum errors are ±6% and ±19%, respectively.

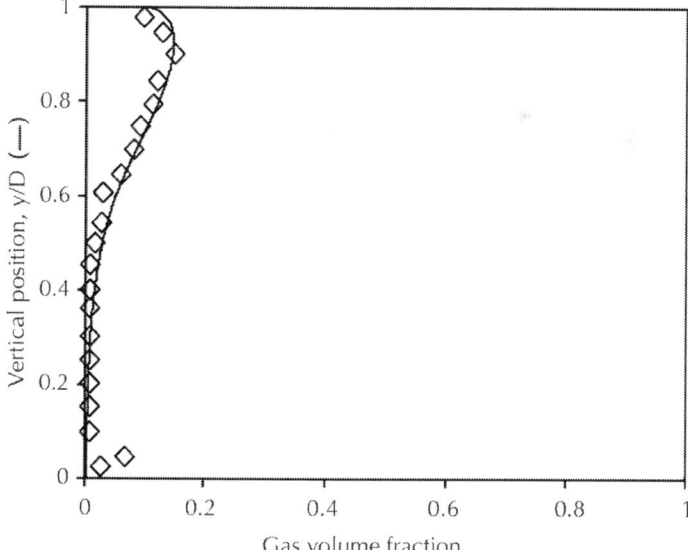

(a)

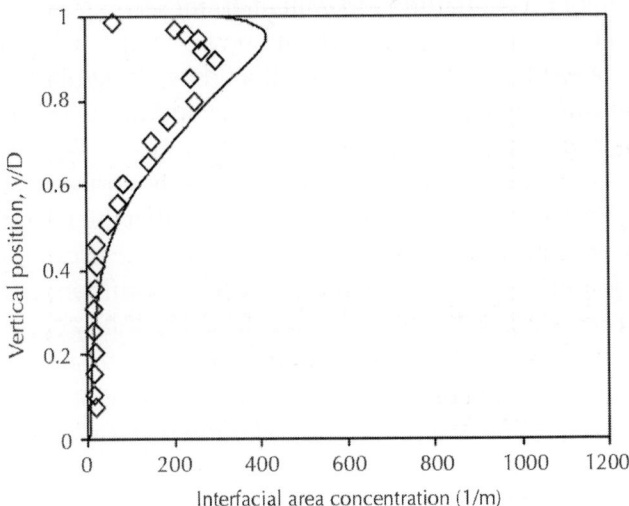

(b)

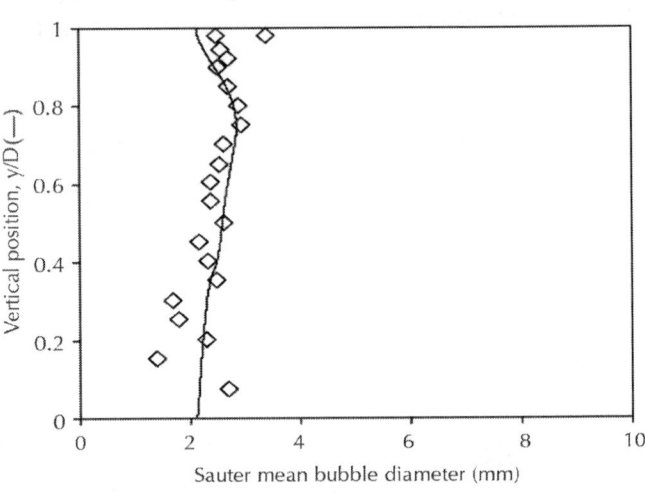

(c)

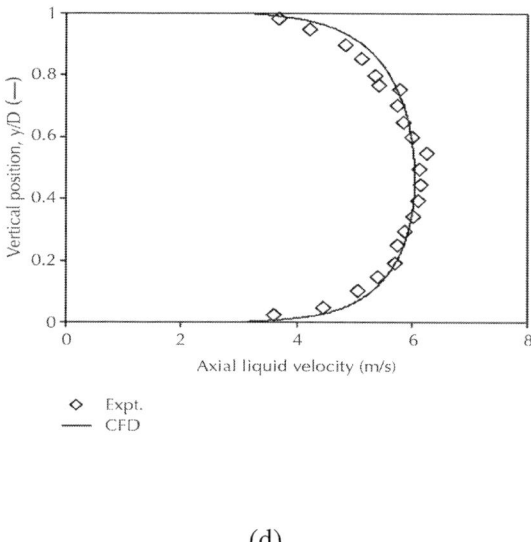

(d)

Figure 3: Comparison of predicted and experimental data of Kocamusta-faogullari and Wang [17] for superficial gas velocity of 0.25 m/s, superficial liquid velocity of 5.1 m/s, and volume fraction 0.043: (a) gas volume fraction, (b) interfacial area concentration, (c) Sauter mean bubble diameter, and (d) axial liquid velocity.

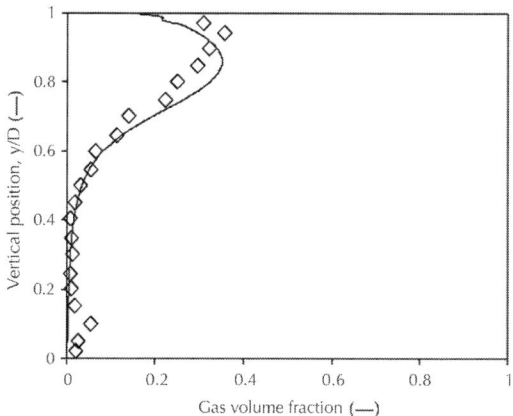

(a)

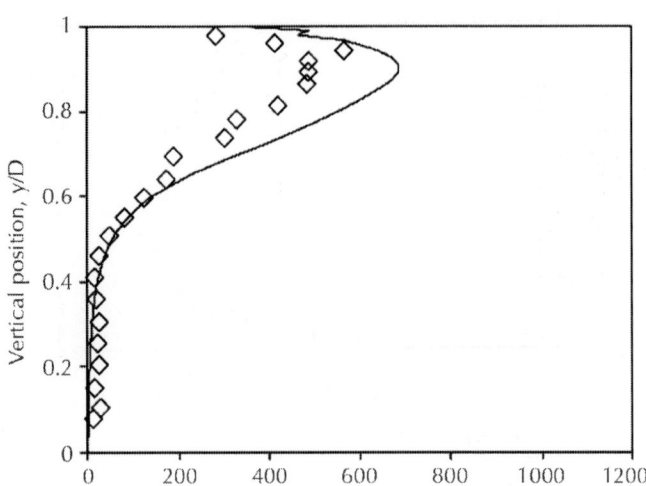

(b)

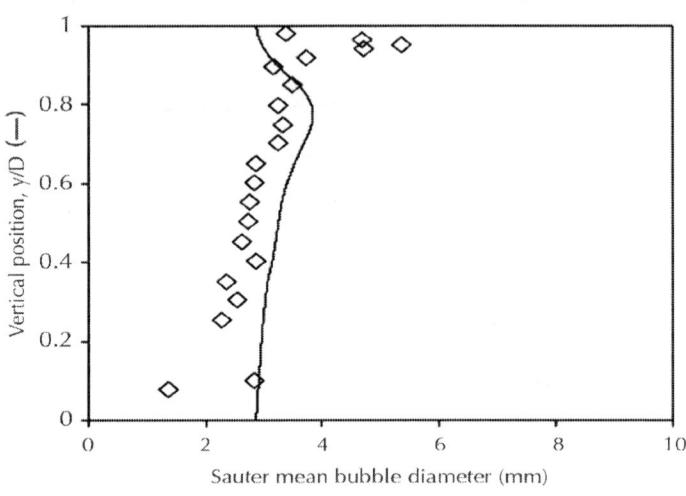

(c)

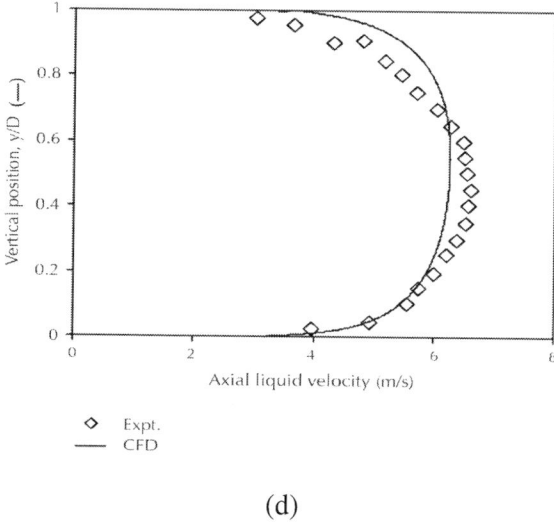

(d)

Figure 4: Comparison of predicted and experimental data of Kocamustafaogullari and Wang [17] for superficial gas velocity of 0.50 m/s, superficial liquid velocity of 5.1 m/s, and volume fraction of 0.080; (a) gas volume fraction, (b) interfacial area concentration, (c) Sauter mean bubble diameter, and (d) axial liquid velocity.

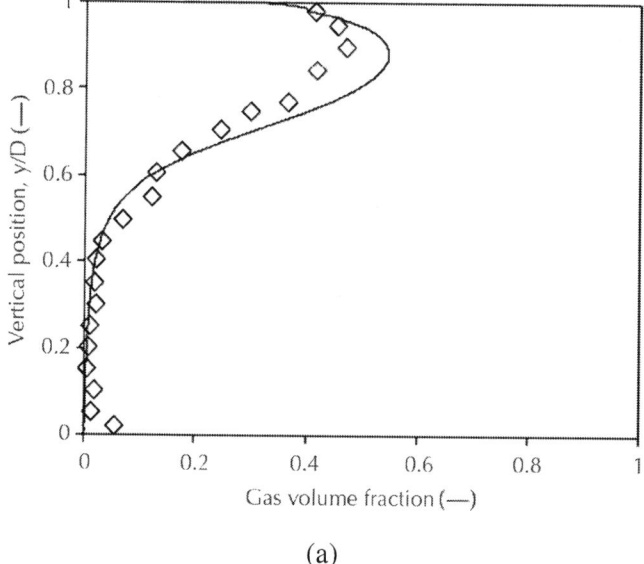

(a)

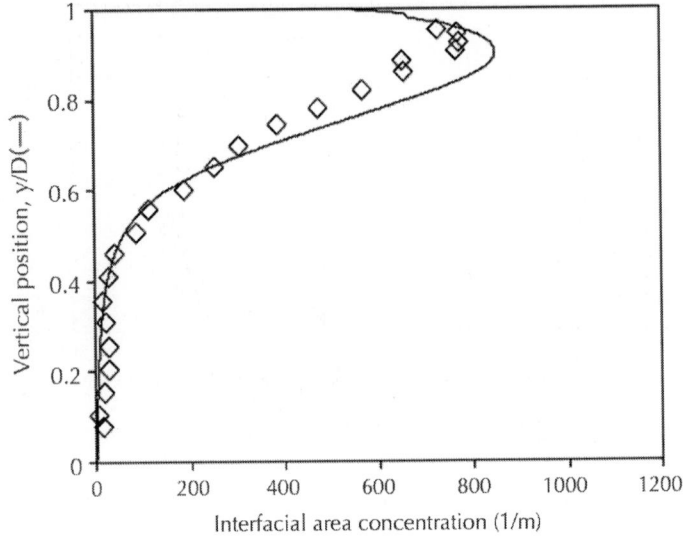

(b)

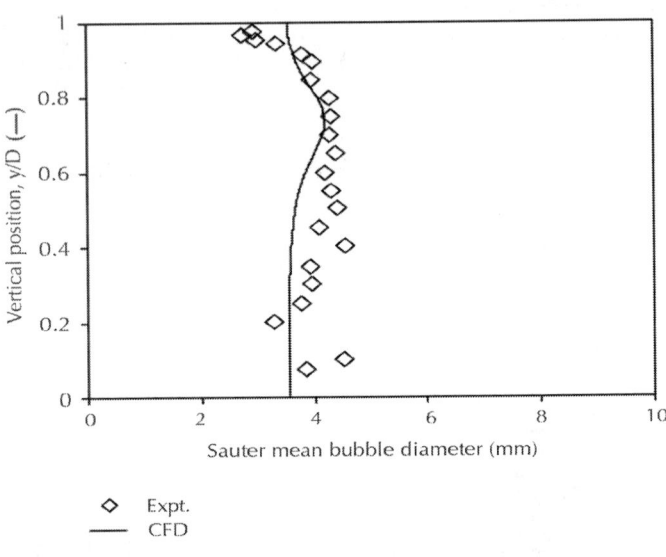

(c)

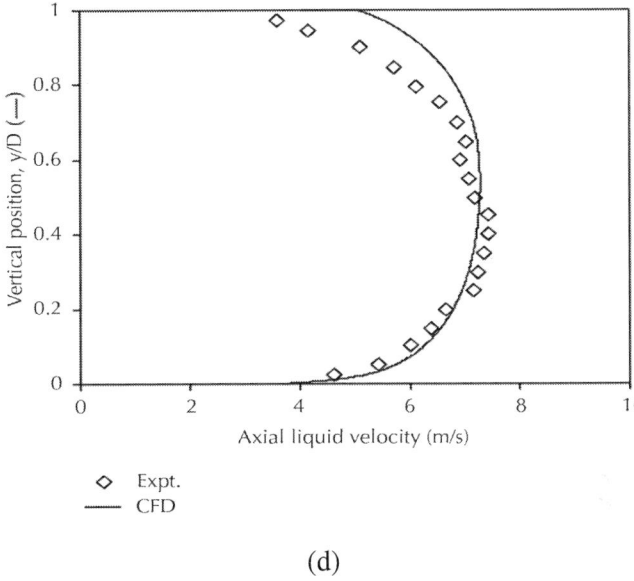

(d)

Figure 5: Comparison of predicted and experimental data of Kocamusta-faogullari and Wang [17] for superficial gas velocity of 0.80 m/s, superficial liquid velocity of 5.1 m/s, and volume fraction of 0.139; (a) gas volume fraction, (b) interfacial area concentration, (c) Sauter mean bubble diameter, and (d) axial liquid velocity.

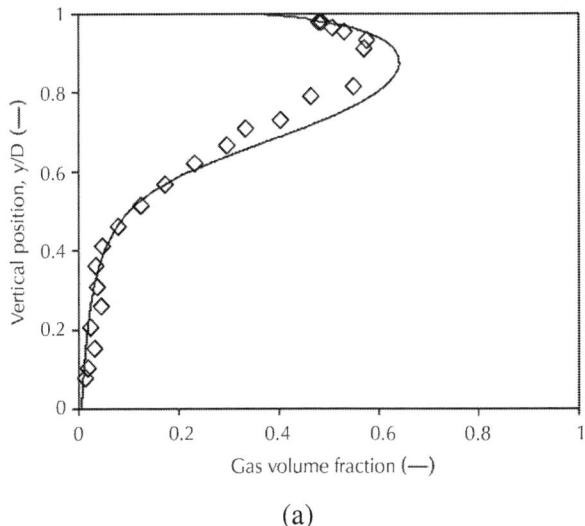

(a)

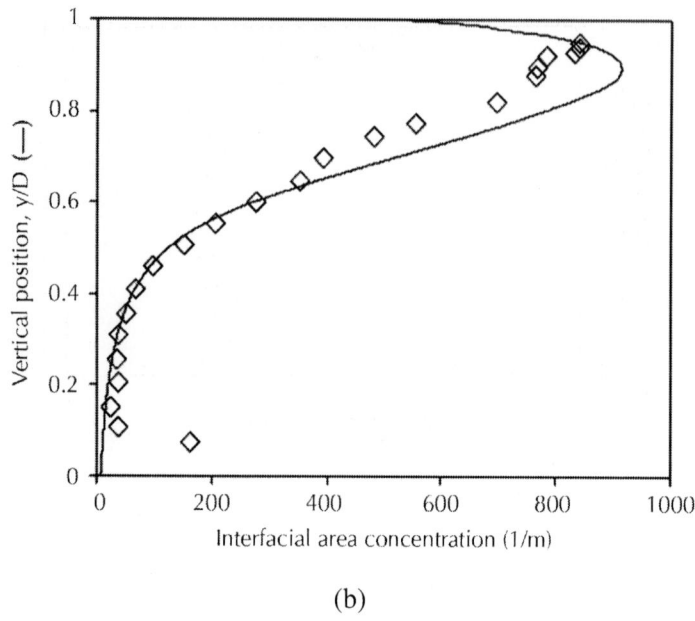

(b)

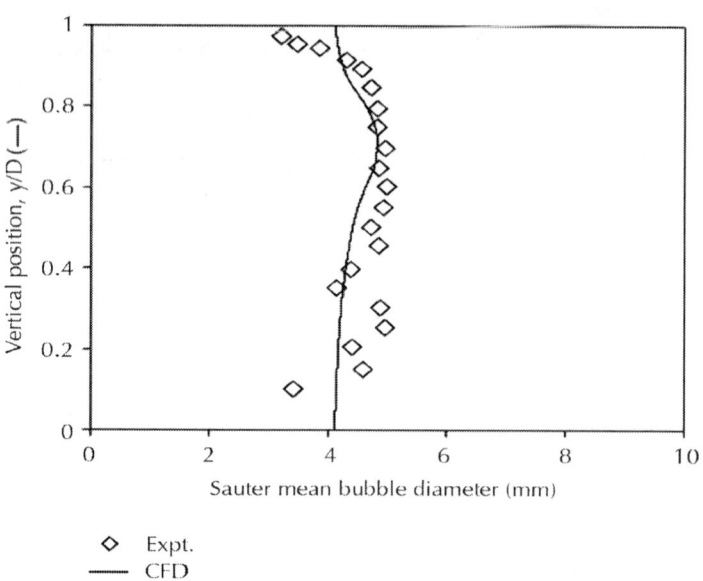

◇   Expt.
——   CFD

(c)

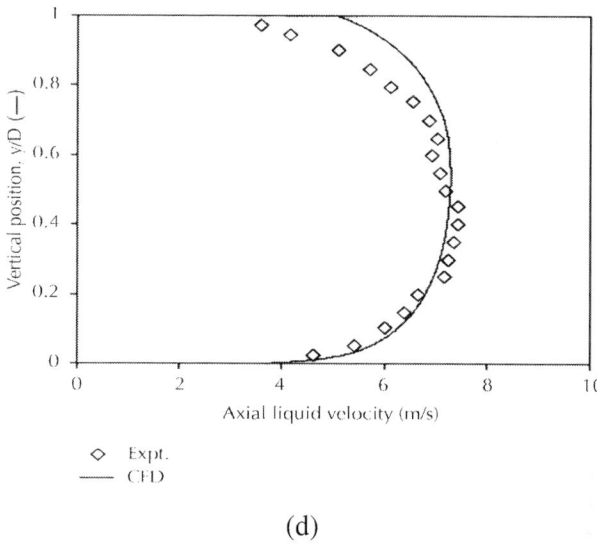

(d)

Figure 6: Comparison of predicted and experimental data of Kocamusta-faogullari and Wang [17] for superficial gas velocity of 1.34 m/s, superficial liquid velocity of 5.1 m/s, and volume fraction of 0.204; (a) gas volume fraction, (b) interfacial area concentration, (c) Sauter mean bubble diameter, and (d) axial liquid velocity.

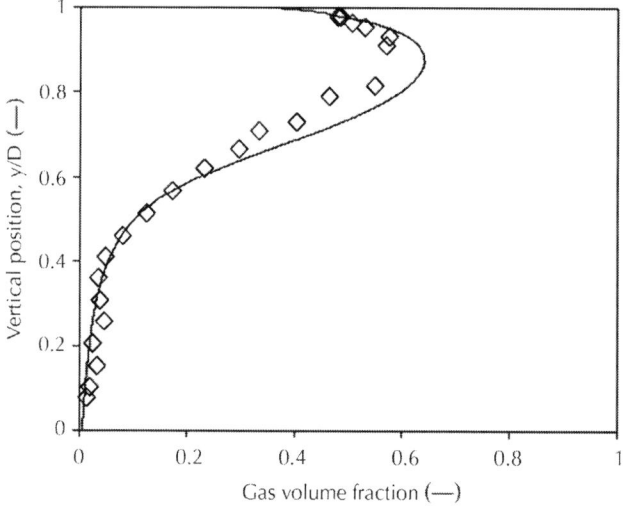

(a)

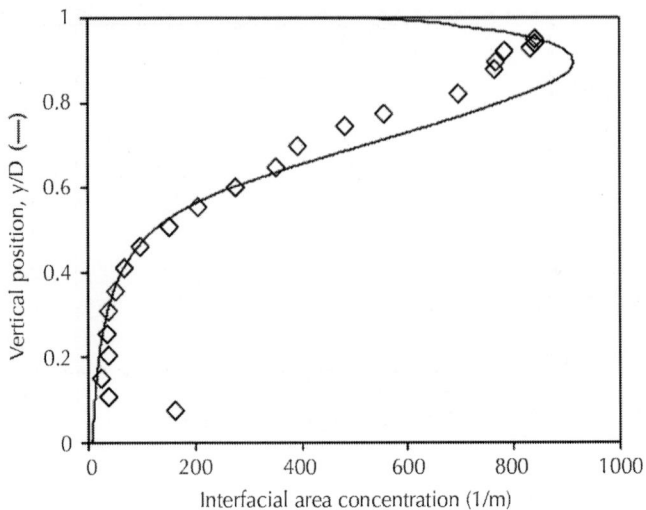

(b)

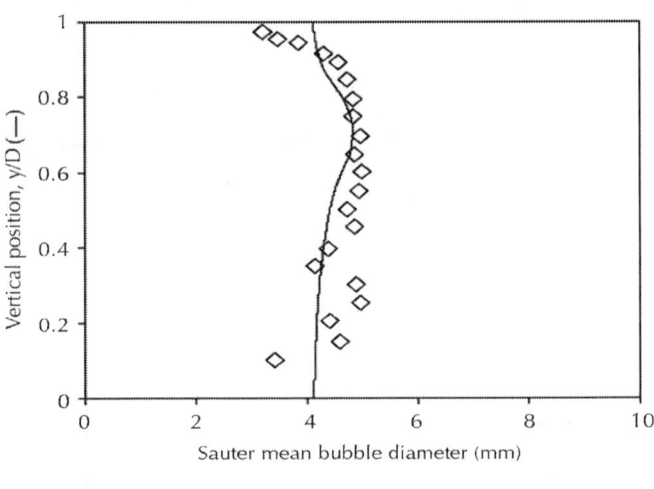

(c)

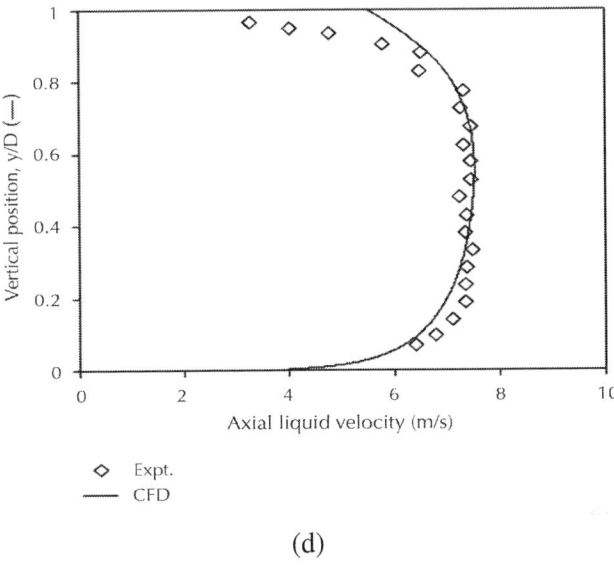

(d)

Figure 7: Comparison of predicted and experimental data of Kocamusta-faogullari and Wang [17] for superficial gas velocity of 0.51 m/s, superficial liquid velocity of 3.74 m/s, and volume fraction of 0.105: (a) gas volume fraction, (b) interfacial area concentration, (c) Sauter mean bubble diameter, and (d) axial liquid velocity.

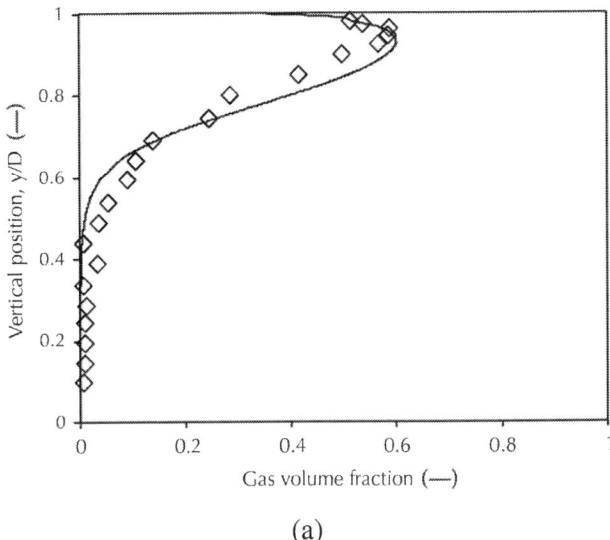

(a)

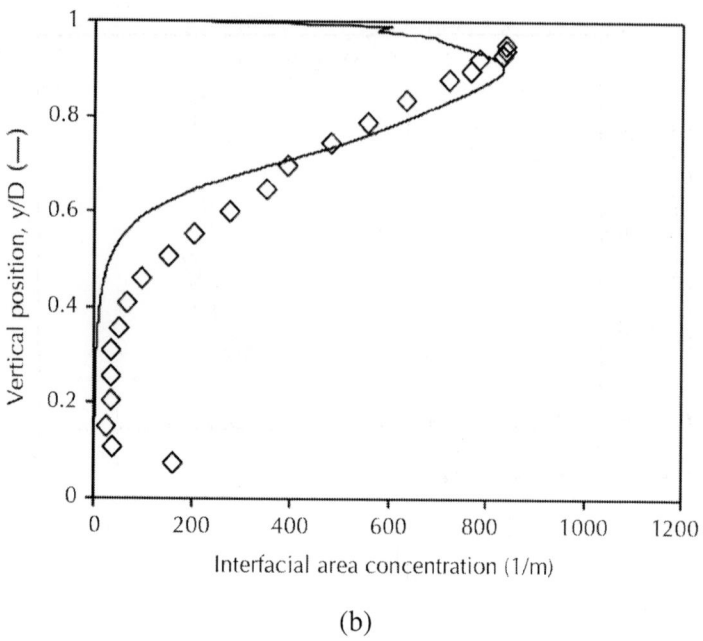

(b)

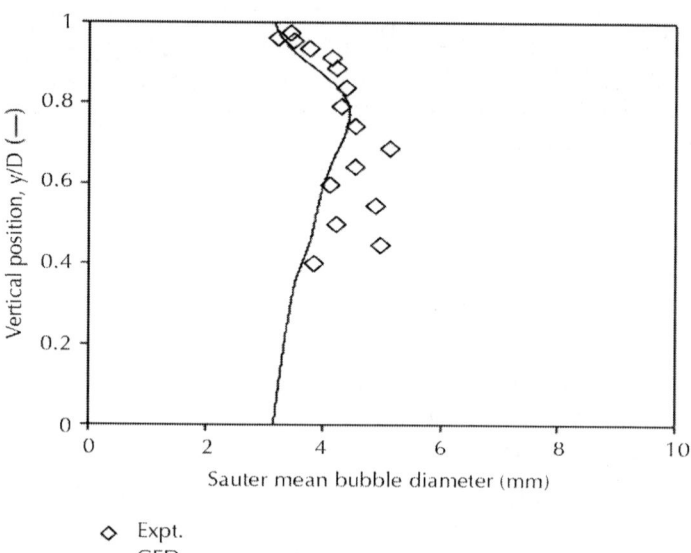

(c)

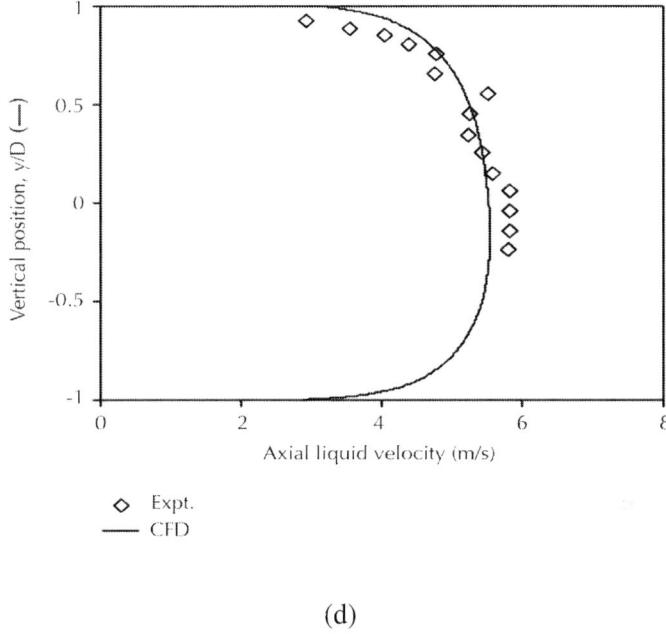

(d)

Figure 8: Comparison of predicted and experimental data of Kocamusta-faogullari and Wang [17] for superficial gas velocity of 0.51 m/s, superficial liquid velocity of 4.4 m/s, and volume fraction of 0.102; (a) gas volume fraction, (b) interfacial area concentration, (c) Sauter mean bubble diameter, and (d) axial liquid velocity.

Interfacial Area Concentration (IAC)

The current simulation results and the experimental results of Kocamustafaogullari and Wang [17] on the local interfacial area concentration variation along the vertical direction are compared in Figures 3(b)–8(b). The flow conditions remain the same in the previous section. The CFD prediction shows good agreement with experimental data. From these figures, it can be seen that the interfacial area concentration shows characteristics that are similar to the gas volume fraction distribution. But the interfacial area depends not only on the volume fraction of the phase, but also on the bubble size distribution. Since the volume fraction and the interfacial area are independent measurements, the data on the interfacial area variation along the vertical direction provide a valuable test of the model predictions from the population balance models. Thus, the agreement seen with the gas volume fraction variation in the previous section provides a level of confidence in the lift coefficient model, while the agreement seen on the interfacial area measurements in the current section provides a level of comfort that the birth and death processes modeled in the population balance model are adequate to describe the bubble dynamics. Further, it can be seen that the local interfacial area concentration can be as high as $1000 \, m^2/m^3$ towards the top of the pipe in horizontal two-phase flow. These values are quite high compared to vertical bubbly flows. This will result in increasing the intensity of the interfacial transport of mass, momentum, and heat near the top of the pipe. In addition, it can be observed that increasing the superficial gas velocity or decreasing the superficial liquid velocity would increase the local and overall interfacial area concentration and tend to flatten the interfacial area concentration profile. The model prediction of interfacial area concentration shows relative mean and maximum errors are ±8% and ±22%, respectively.

Sauter Mean Bubble Diameter

The comparison of predicted and experimental data of the local Sauter mean bubble diameter distribution is shown in Figures 3(c)–8(c) for various superficial gas and liquid velocities. The Sauter mean bubble diameters are in the range of 1.5–5 mm, depending on the location and flow conditions. It should be noted that the experimental data on Sauter mean diameter is inferred from other measurements, and it is not a directly measured quantity. The scatter in the experimentally derived data is high, particularly in the lower region where the gas volume fraction is low, indicating that perhaps the signals are weaker in that region. Good agreement was achieved against the measured bubble size for all the experimental conditions. From these figures, it can be seen that the bubble size distribution is almost uniform in the pipe cross-section except near the wall region. The Suater mean bubble size tends to reduce close to the top of the pipe wall. This can be attributed to the fact that near the wall a very strong velocity gradient exists, which causes further break-up into smaller bubble sizes. Furthermore, the Suater mean bubble size is seen to increase with increasing the superficial gas velocity (Figures 3(c)–8(c)) and to decrease with increasing superficial liquid velocity (Figures7(c)–8(c)). The simulation results capture all of these trends faithfully. The model prediction shows relative mean and maximum errors are ±9% and ±24%, respectively.

Axial Liquid Velocity

Figures 3(d)–8(d) show the comparison of predicted and experimental data of axial liquid velocity profiles for different superficial gas and liquid velocities. If only a single liquid phase moves in the pipe, the liquid velocity in the pipe top region will be equal to the velocity in the bottom region, exhibiting a perfect axisymmetry. But these results show that the axial liquid velocity profile has a slight degree of asymmetry due to the presence of gas flow. The degree of asymmetry decreases with increasing liquid flow or decreasing gas flow. For increasingly higher gas velocities (Figures 3(d)–6(d)), the liquid velocity in the upper region of the pipe is slightly lower than in the lower region. This could be attributed to larger volume fraction of gas in the upper region which is the reason for the asymmetric distribution of the liquid velocity. An interesting feature of the velocity profile is that the velocity distribution within the bottom liquid layer resembles closely a fullydeveloped turbulent pipe flow profile irrespective of the liquid and gas superficial velocities. The model prediction of axial liquid velocity shows relative mean and maximum errors are ±5% and ±14%, respectively.

Simulation Results

From the simulation, we can get much more additional information, while some of these quantities are more difficult to measure in an experiment. One such quantity is the slip velocity between the two phases. The variation in the vertical direction of the slip velocity is shown in Figure 9 for various combinations of gas and liquid flow rates. The slip velocity is larger in magnitude near the top region of the pipe, while a smaller slip velocity exists in the bottom part of the pipe. The slip velocity is an important characteristic of two-phase flow, particularly because of the large difference in densities between phases. Relatively smaller bubbles and fewer in number are found in the bottom region, and hence, they tend to move with the liquid resulting in a smaller slip velocity, while relatively larger bubbles and more in number are found near the top of the pipe, resulting in a larger slip velocity.

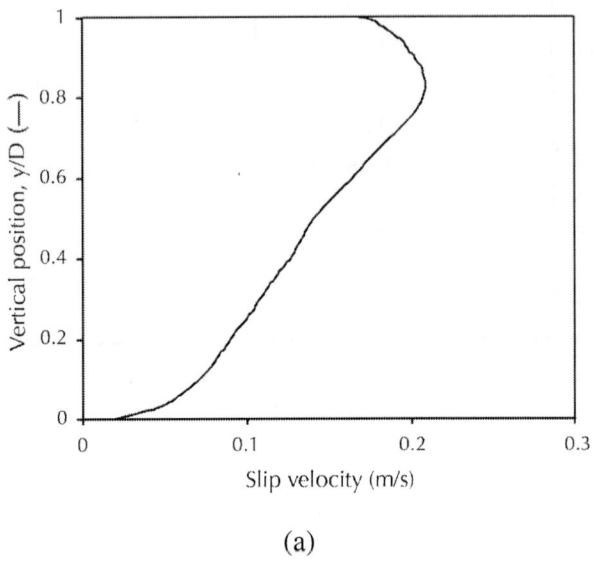

(a)

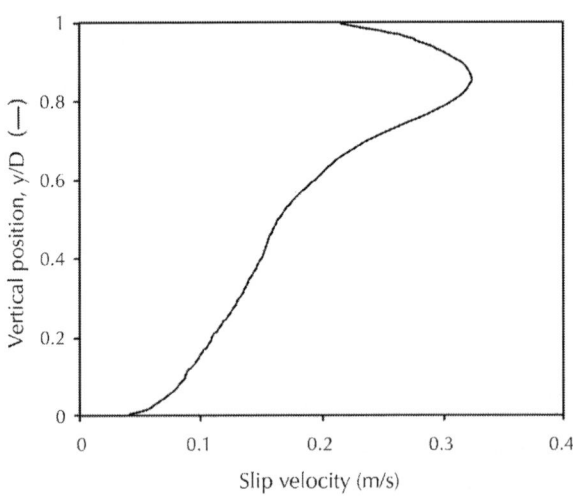

(b)

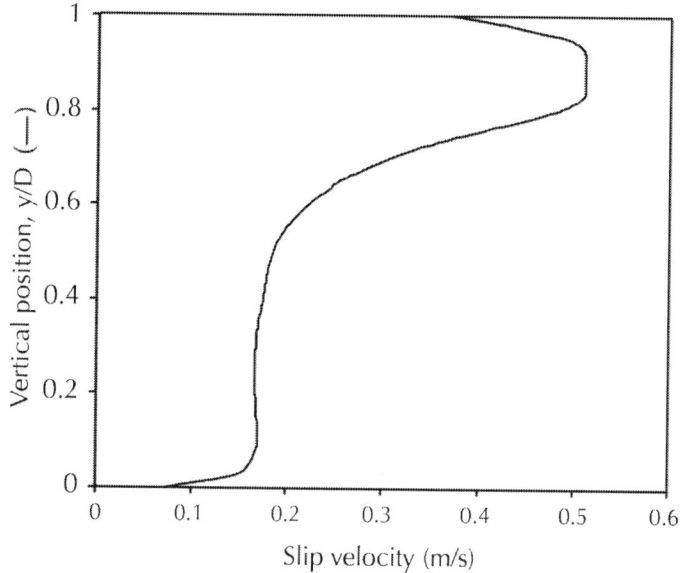

(c)

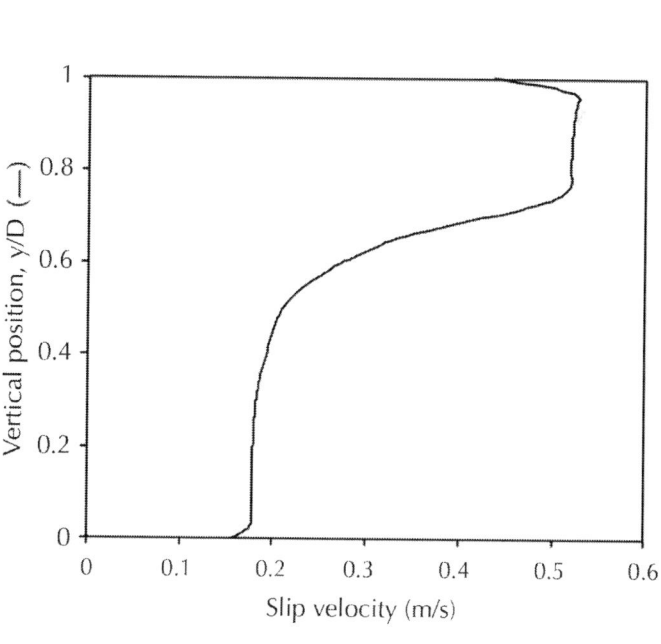

(d)

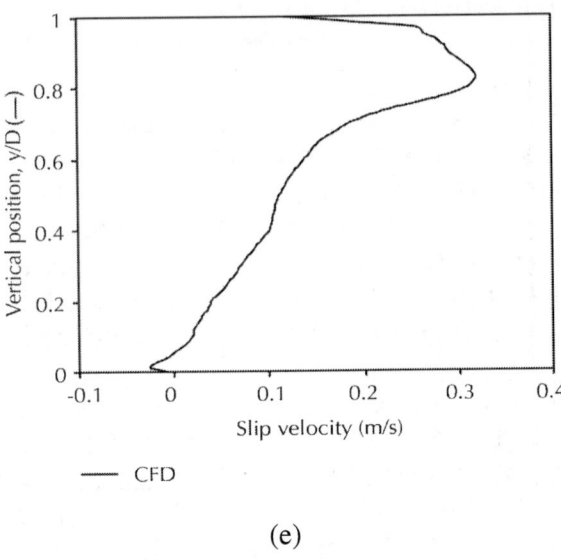

(e)

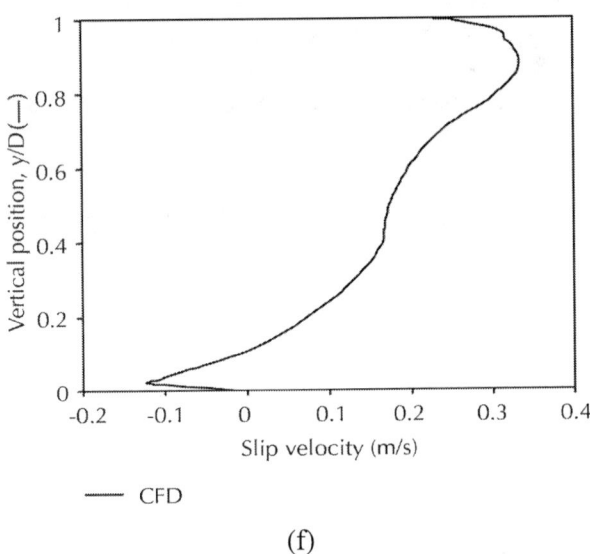

(f)

Figure 9: Slip velocity at different superficial gas and liquid velocities: (a) V_G = 0.25 m/s, V_L = 5.1 m/s; (b) V_G = 0.50 m/s, V_L = 5.1 m/s; (c) V_G = 0.80 m/s, V_L = 5.1 m/s; (d) V_G = 1.34 m/s, V_L = 5.1 m/s; (e) V_G = 0.51 m/s, V_L = 3.74 m/s; (f) V_G = 0.51 m/s, V_L = 4.4 m/s.

Average Interfacial Parameters

While we have used the data on the spatial variation of quantities such as volume fraction and interfacial area to tune the model parameters, from a practical view point one is often interested only in a quantity that is averaged over the pipe cross-section. Hence, area averaged gas volume fraction, interfacial area concentration, and mean bubble diameter at the exit plane are shown in Figure 10 as a function of superficial gas velocity at various liquid velocities of 5.1 m/s, 4.4 m/s, and 3.74 m/s. The average volume fraction and interfacial area increase significantly with increasing superficial gas velocity, as expected. The influence of superficial liquid velocity on the gas volume fraction and interfacial area concentration are less significant. Figure 10(c) shows that the average bubble diameter increases slightly with increasing superficial gas velocity, all though the influence is not significant. However, the mean bubble size decreases with increasing superficial liquid velocity. This observation supports the fact that the bubble size is determined primarily by liquid flow turbulence in horizontal flows. Figure 10 compares the measured gas volume fraction, interfacial area, and Sauter bubble mean diameter values with those predicted using CFD-PBM model, and the relative mean and maximum errors are ±4% and ±11%, respectively.

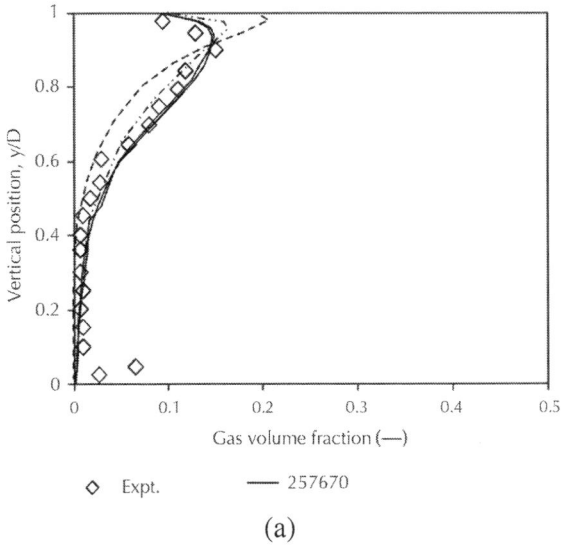

(a)

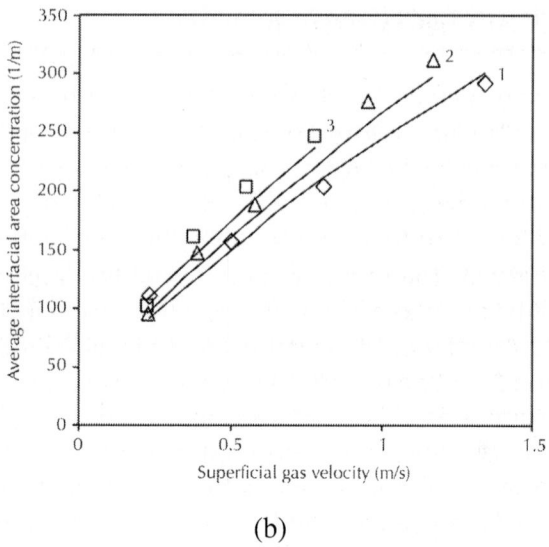

(b)

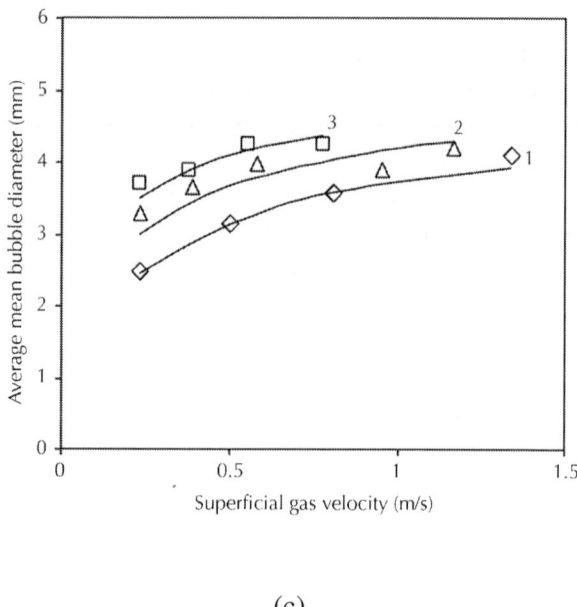

(c)

Figure 10: Effect of superficial gas and liquid velocity on (a) average gas volume fraction; (b) average interfacial area concentration (c) average mean bubble diameter: (1) $V_L = 5.1$m/s; (2) $V_L = 4.4$m/s; (3) $V_L = 3.74$ m/s.

Bubble Size Distribution

The bubble size distribution is determined by bubble coalescence and breakup. In a given system, bubble coalescence and breakup are primarily influenced by the local gas volume fraction and kinetic energy dissipation rate. Because of the nonuniform profiles of the gas volume fraction and dissipation rate, the bubble size distribution varies with the position as well. The spatial evolution of bubble size distribution between the inlet and the outlet of the pipe is shown in Figure 11 for a superficial gas velocity of 0.25 m/s and a superficial liquid velocity of 4.67 m/s. While selecting the bubble size, we have considered a range from 1 to 10 mm, and the experimental observation of bubble size for highest superficial velocity is 6 mm. Figures 11(a), 11(c), and11(e) show the bubble size distribution that was specified at the pipe inlet. These correspond, respectively, to the monosized bubbles of 1.45 mm (Figure 11(a)), 9.55 mm (Figure 11(c)), and a uniform distribution of bubbles in the range of 1 to 10 mm (Figure 11(e)). The corresponding distribution at the pipe exit is shown on the right hand side in Figures 11(b), 11(d), and 11(f), respectively. It can be seen from these figures that the bubble size distribution function reaches an independent state as determined by the balance between birth and death processes that depend on the local flow conditions, and its original state at the inlet has very little impact. Although this kind of distribution function was not measured in the experiments of Kocamustafaogullari and Wang [17], the spatial variation of the bubble sizes was measured as shown in Figure 3. It is comforting to note that the ranges of bubbles sizes measured under similar flow conditions show a range of 2-3 mm, the same range shown in Figure 11, even though extremely small (1.45 mm) and large (9.55 mm) sizes were used at the pipe inlet.

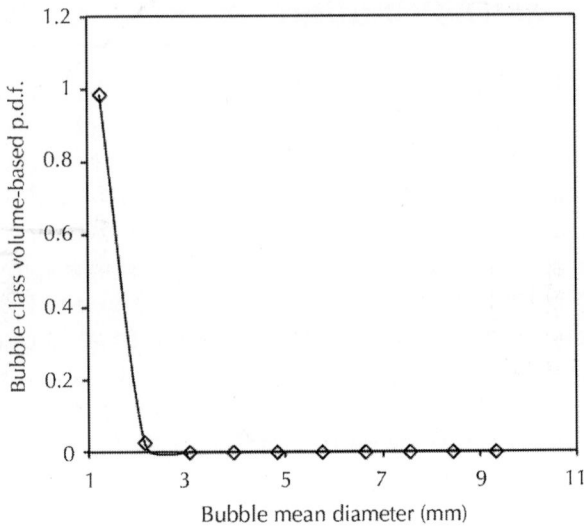

(a)

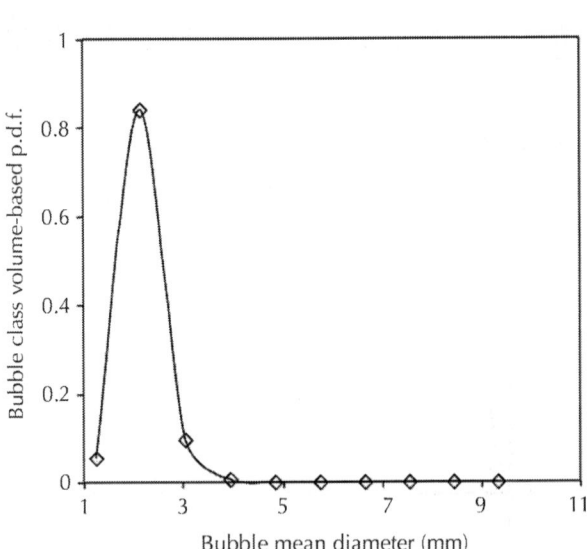

(b)

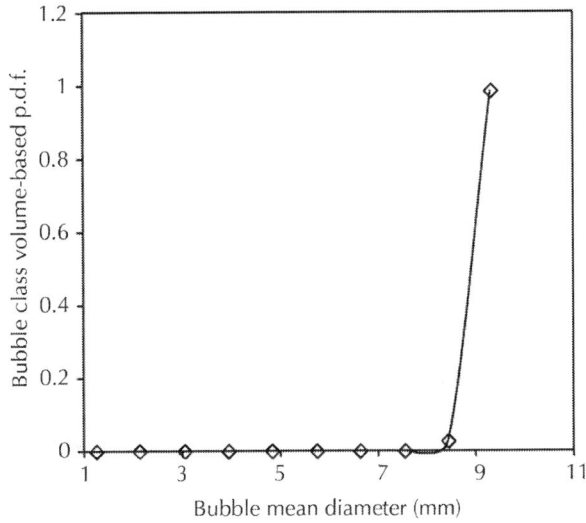

(c)

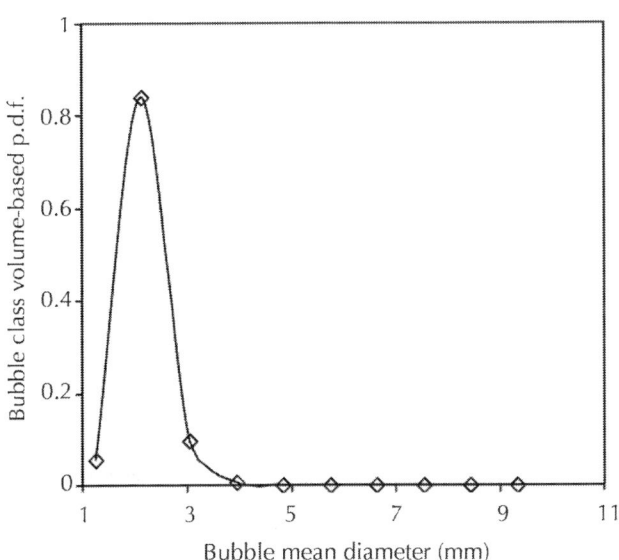

(d)

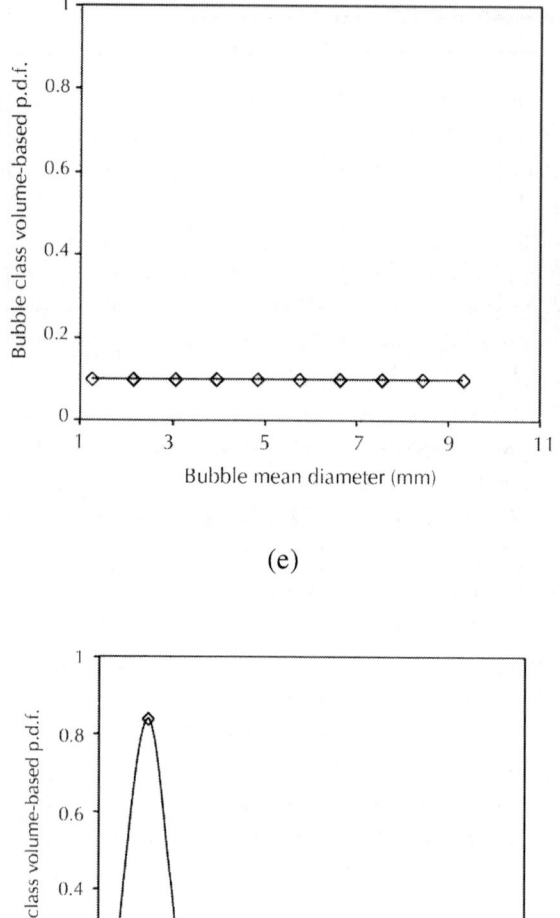

(e)

(f)

Figure 11: The bubble class volume-based p.d.f at inlet and exit of the pipe for superficial gas velocity is 0.25 m/s, superficial liquid velocity is 4.67 m/s, and average volume fraction is 0.043.

Development of Flow Pattern

To see the development of flow pattern in the axial direction, several three-dimensional simulations were carried out using the coupled two-fluid and population balance models. The flow evolution is shown in Figure12–14 at three different axial locations of L/D = 25, 148, and 253. The first location represents close to the entrance of the pipe region where the internal flow develops, and the second and third locations indicate the extent to which the flow has reached a fully developed state, by the lack of further change in flow profiles. Figures 12, 13, and 14 show, respectively, the development of the local gas volume fraction, interfacial area concentration, and axial liquid velocity in axial direction for the superficial gas velocity of 1.21 m/s and the superficial liquid velocity of 4.67 m/s. Good agreement can be seen between the predicted and experimental data at the axial location of L/D = 25. The gas volume fraction and interfacial area concentration do not show a significant variation in the vertical direction, near the entrance of the pipe (L/D = 25). This is because the bubble residence time was very small, and the transverse phase segregation due to the gravity has not been established yet. However, from first location (L/D = 25) to the second location (L/D = 148), the large differences can be observed. From second location (L/D = 148) to third location (L/D = 253), there is no significant difference was observed, but the fluid segregation due to the buoyancy is still effective. Further, it can be observed from Figure 14 that the axial liquid velocity profile shows nearly the same for all the locations. A slight change in the numerical values of the velocity can be attributed to the expansion of the gas phase associated with the frictional pressure gradient causing a continuous acceleration of the mixture in the axial direction.

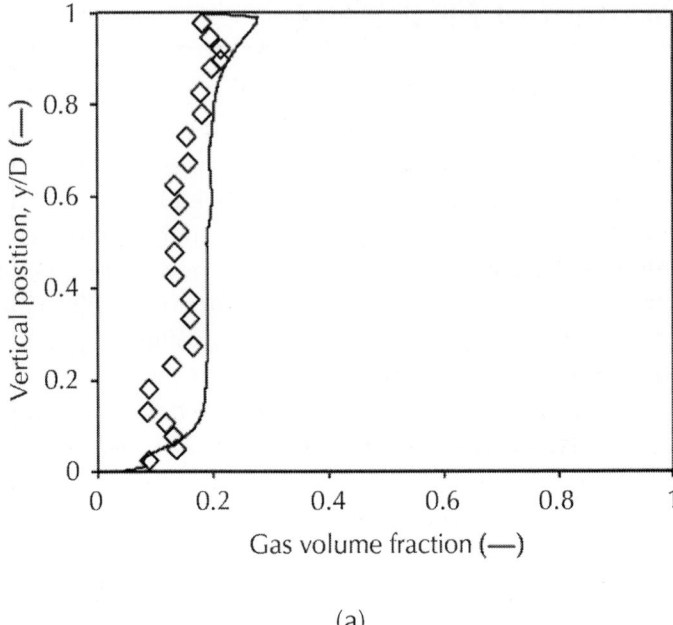

(a)

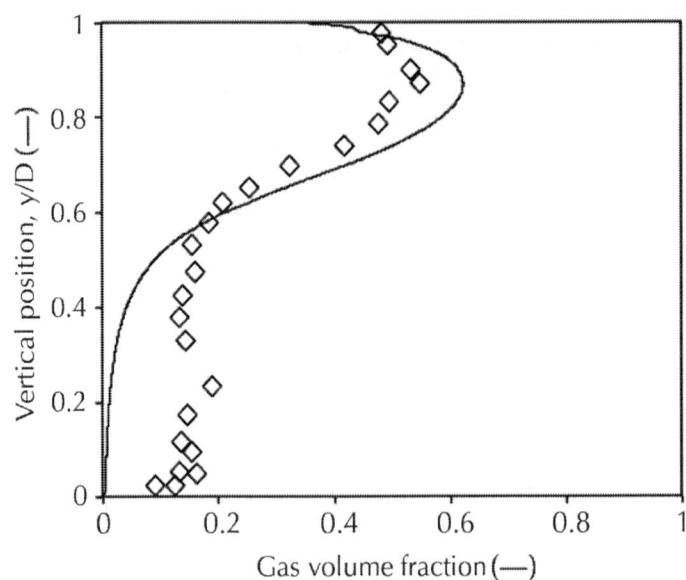

(b)

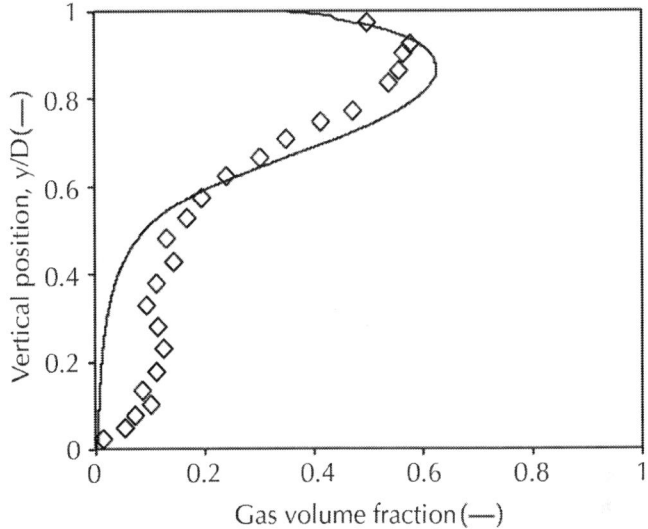

(c)

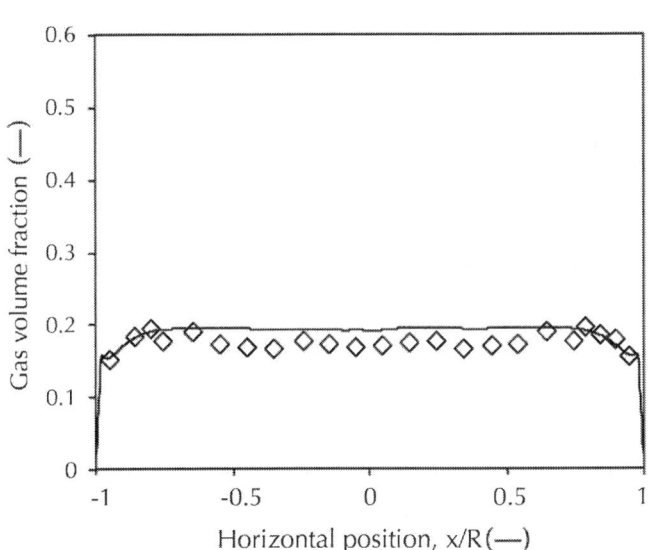

(d)

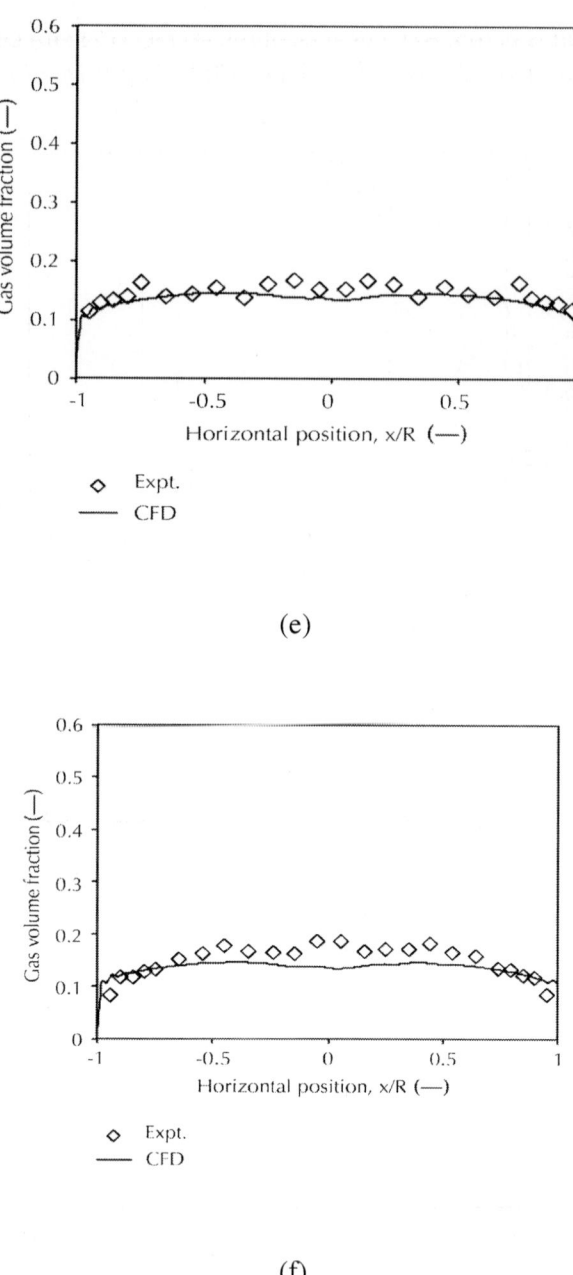

(e)

(f)

Figure 12: Gas volume fraction development in axial direction for superficial gas velocity is 1.21 m/s, superficial liquid velocity is 4.67 m/s, and average volume fraction is 0.205. At vertical position, (a) L/D = 25; (b) L/D = 148; (c) L/D = 253, at horizontal position, (d) L/D= 25; (e) L/D = 148; (f) L/D = 253.

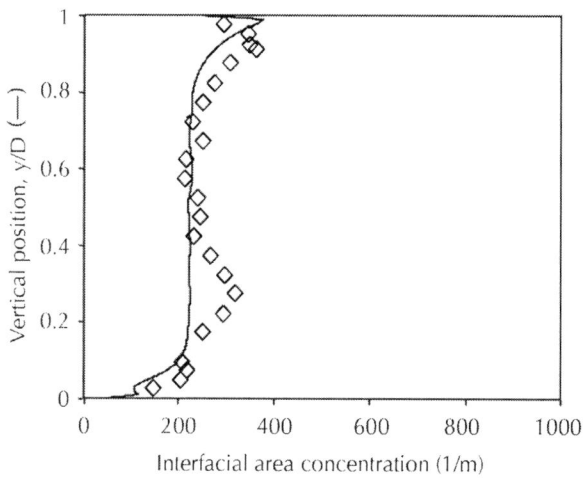

(a)

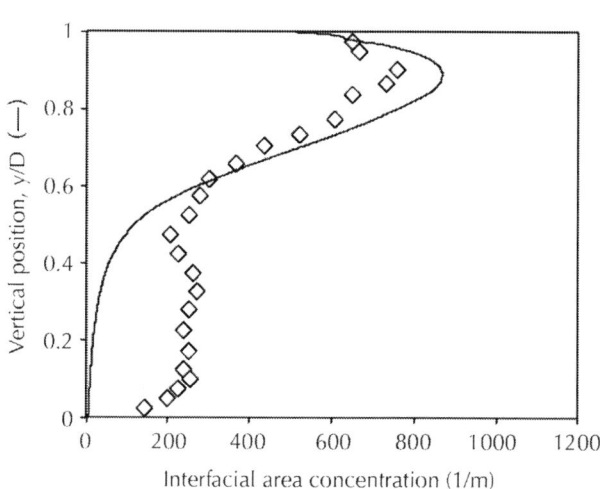

(b)

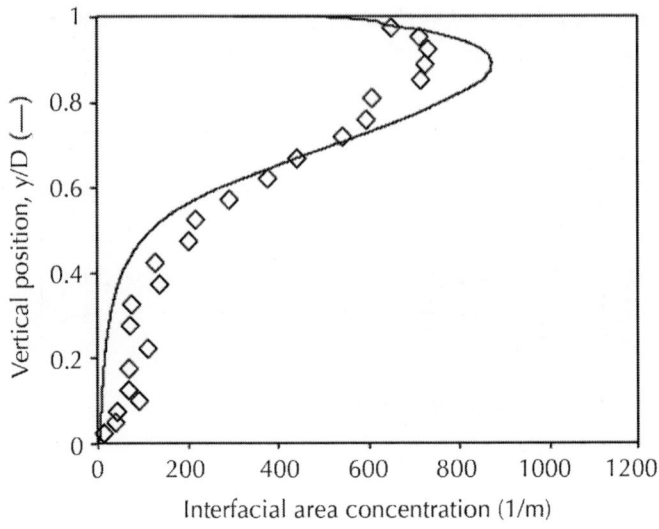

(c)

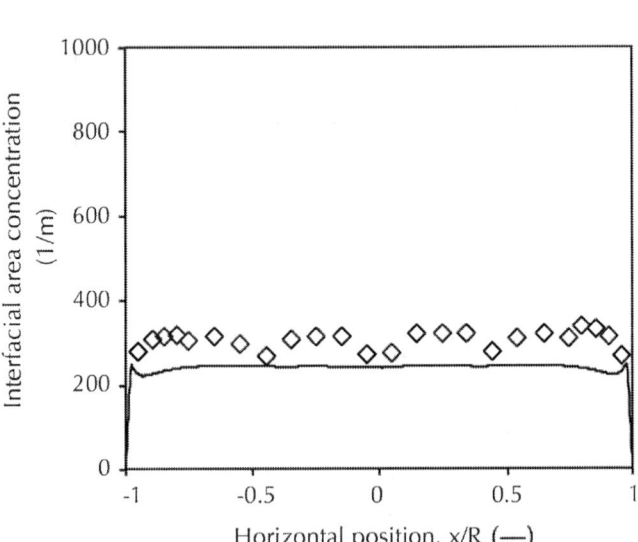

(d)

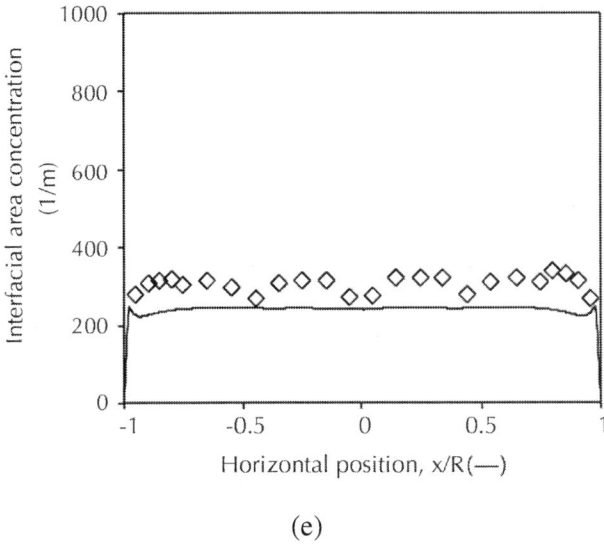

(e)

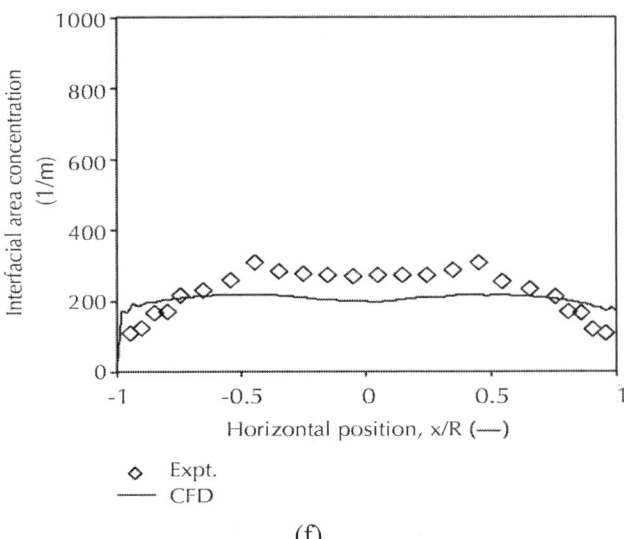

(f)

Figure 13: Interfacial area concentration (IAC) development in axial direction for superficial gas velocity is 1.21 m/s, superficial liquid velocity is 4.67 m/s, and average volume fraction is 0.205. At vertical position, (a) L/D = 25; (b) L/D = 148; (c) L/D = 253, at horizontal position, (d) L/D = 25; (e) L/D = 148; (f) L/D = 253.

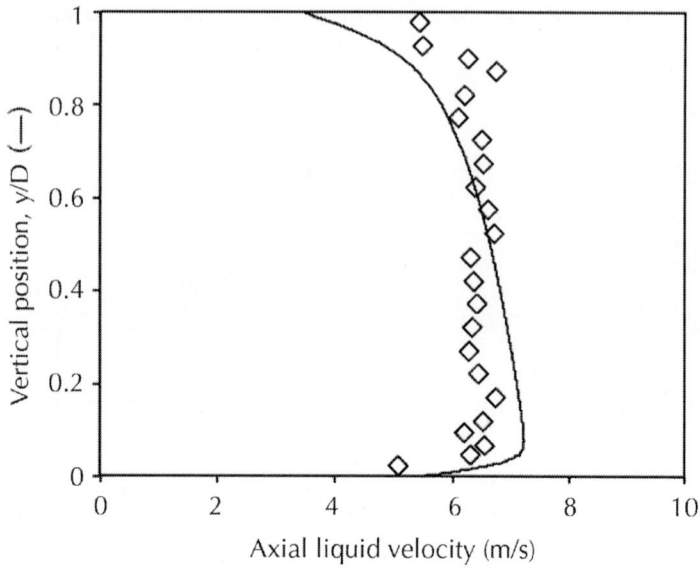

(a)

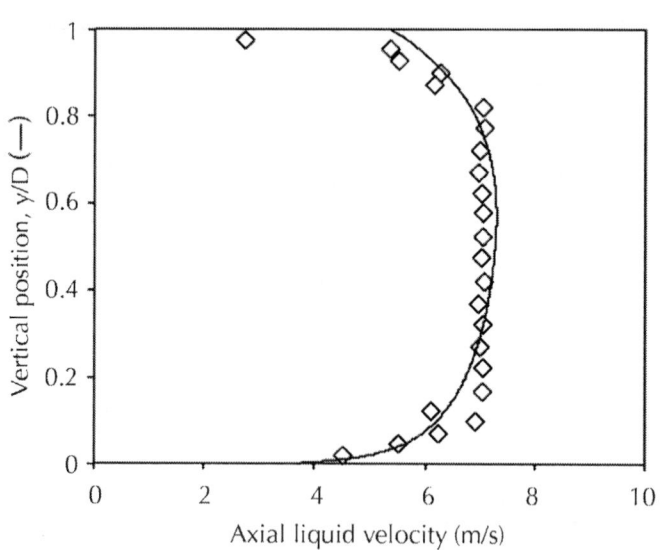

(b)

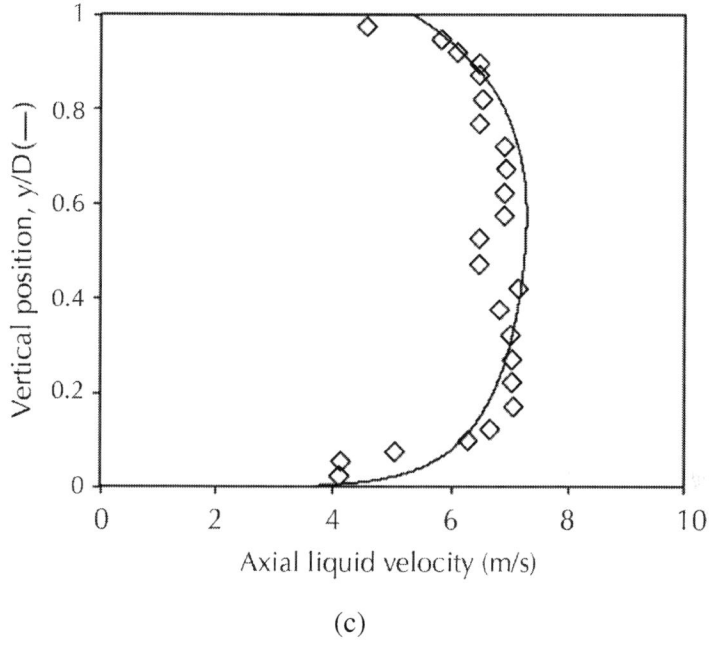

(c)

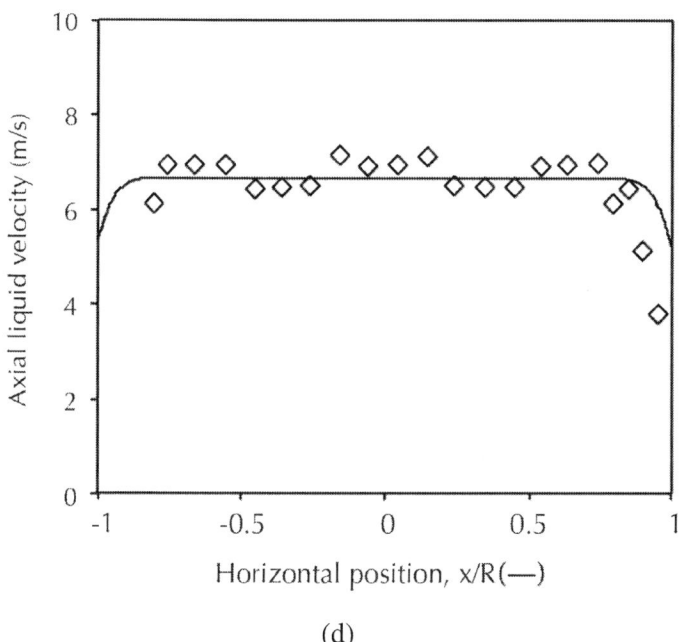

(d)

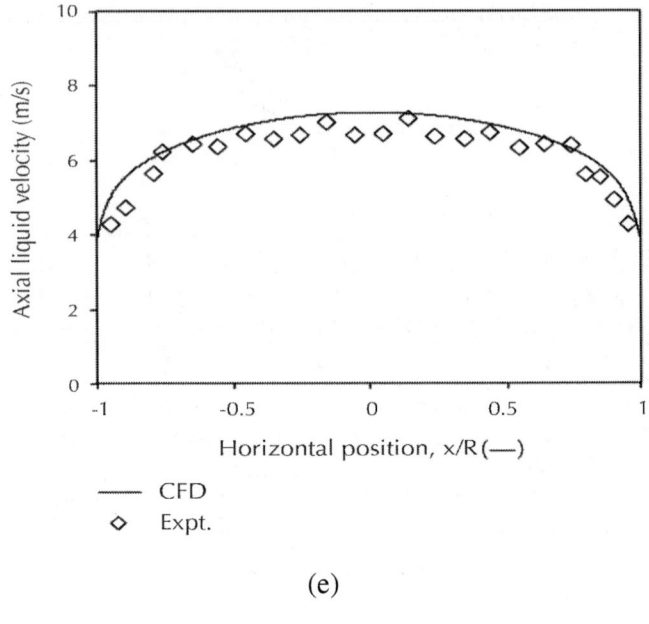

(e)

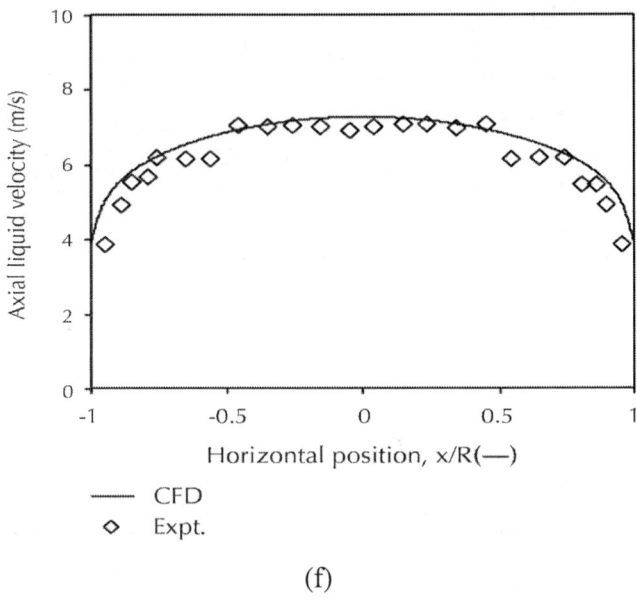

(f)

Figure 14: Liquid velocity development in axial direction for superficial gas velocity is 1.21 m/s, superficial liquid velocity is 4.67 m/s, and average volume fraction is 0.205. At vertical position, (a) L/D = 25; (b) L/D = 148; (c) L/D = 253, at horizontal position, (d) L/D= 25; (e) L/D = 148; (f) L/D = 253.

CONCLUSIONS

A two-fluid model coupled with population balance approach is presented in this paper to handle gas-liquid bubbly flows in horizontal pipe. To demonstrate the application of the population balance approach, the average bubble number density transport equation was formulated and implemented for gas-liquid bubbly flows in the CFD code CFX 5.7 to determine the temporal and spatial geometrical changes of the gas bubbles. Population balance combined with coalescence and break-up models were taken into consideration. A detailed comparison has been presented between the CFD simulation and the experimental data reported by Kocamustafaogullari and Wang [17] and Kocamustafaogullari and Huang [1]. Good agreement was seen between the predicted and the experimental data of the volume fraction, interfacial area concentration, Sauter mean bubble diameter, and liquid velocity for a range of superficial gas (0.25 to 1.34 m/s) and liquid (3.74 to 5.1 m/s) velocities and volume fraction (4 to 21%). The experimental and simulated results indicate that the volume fraction and interfacial area concentration have local maxima near the upper pipe wall, and the profiles tend to flatten with increasing liquid flow rate. It was observed that the mean bubble diameter ranged from 1.5 to 5 mm, depending on the location and flow conditions. Further, it was found that increasing the gas flow rate at fixed liquid flow rate would increase the local volume fraction and interfacial area concentration. The simulation results were consistent with experimental observed from the literature. Further, the development of flow pattern was examined at three axial locations L/D = 25, 148, and 253. It was found that the prediction shows good agreement with experimental data. The axial liquid mean velocity showed a relatively uniform distribution except near the upper pipe wall. The flow in the bottom part of the pipe exhibits a fully developed turbulent pipe flow profile, whereas in the top of the pipe a different flow exists.

ACKNOWLEDGMENTS

The authors gratefully acknowledge the financial support from the Natural Sciences and Engineering Research Council of Canada (NSERC) and Syncrude Canada Ltd. for this project.

REFERENCES

1. G. Kocamustafaogullari and W. D. Huang, "Internal structure and interfacial velocity development for bubbly two-phase flow," Nuclear Engineering and Design, vol. 151, no. 1, pp. 79–101, 1994.

2. R. Krishna and J. M. van Baten, "Scaling up bubble column reactors with the aid of CFD," Chemical Engineering Research and Design, vol. 79, no. 3, pp. 283–309, 2001. · ·

3. T. Hibiki and M. Ishii, "Development of one-group interfacial area transport equation in bubbly flow systems," International Journal of Heat and Mass Transfer, vol. 45, no. 11, pp. 2351–2372, 2002. · ·

4. A. Kitagawa, K. Sugiyama, and Y. Murai, "Experimental detection of bubble-bubble interactions in a wall-sliding bubble swarm," International Journal of Multiphase Flow, vol. 30, no. 10, pp. 1213–1234, 2004. · ·

5. Z. Xiao and R. B. H. Tan, "A model for bubble-bubble and bubble-wall interaction in bubble formation,"AIChE Journal, vol. 52, no. 1, pp. 86–98, 2006. · ·

6. D. Colella, D. Vinci, R. Bagatin, M. Masi, and E. Abu Bakr, "A study on coalescence and breakage mechanisms in three different bubble columns," Chemical Engineering Science, vol. 54, no. 21, pp. 4767–4777, 1999.

7. E. Olmos, C. Gentric, Ch. Vial, G. Wild, and N. Midoux, "Numerical simulation of multiphase flow in bubble column reactors. Influence of bubble coalescence and break-up," Chemical Engineering Science, vol. 56, no. 21-22, pp. 6359–6365, 2001.

8.   F. Lehr, M. Millies, and D. Mewes, "Bubble-size distributions and flow fields in bubble columns," AIChE Journal, vol. 48, no. 11, pp. 2426–2443, 2002. · ·

9.   F. B. Campos and P. L. C. Lage, "A numerical method for solving the transient multidimensional population balance equation using an Euler-Lagrange formulation," Chemical Engineering Science, vol. 58, no. 12, pp. 2725–2744, 2003.

10. P. Chen, J. Sanyal, and M. P. Dudukovic, "CFD modeling of bubble columns flows: implementation of population balance," Chemical Engineering Science, vol. 59, no. 22-23, pp. 5201–5207, 2004. · ·

11. J. Sanyal, D. L. Marchisio, O. Fox, and K. Dhanasekharan, "On the comparison between population balance models for CFD simulation of bubble columns," Industrial and Engineering Chemistry Research, vol. 44, no. 14, pp. 5063–5072, 2005. · ·

12. H. A. Jakobsen, H. Lindborg, and C. A. Dorao, "Modeling of bubble column reactors: progress and limitations," Industrial and Engineering Chemistry Research, vol. 44, no. 14, pp. 5107–5151, 2005. · ·

13. Z. Sha, A. Laari, and I. Turunen, "Multi-phase-multi-size group model for the inclusion of population balances into the CFD simulation of gas-liquid bubbly flows," Chemical Engineering and Technology, vol. 29, no. 5, pp. 550–559, 2006. ·

14. T. Wang, J. Wang, and Y. Jin, "A CFD-PBM coupled model for gas-liquid flows," AIChE Journal, vol. 52, no. 1, pp. 125–140, 2006. · ·

15. S. Lo, "Application of the MUSIG model to bubbly flows," AEAT -1096, AEA Technology, 1996.

16. G. H. Yeoh and J. Y. Tu, "Population balance modelling for bubbly flows with heat and mass transfer,"Chemical Engineering Science, vol. 59, no. 15, pp. 3125–3139, 2004. · ·

17. G. Kocamustafaogullari and Z. Wang, "An experimental study on local interfacial parameters in a horizontal bubbly two-phase flow," International Journal of Multiphase Flow, vol. 17, no. 5, pp. 553–572, 1991.

18. H. Luo and H. F. Svendsen, "Theoretical model for drop and bubble break-up in turbulent dispersions,"AIChE Journal, vol. 42, no. 5, pp. 1225–1233, 1996.

19. M. J. Prince and H. W. Blanch, "Bubble coalescence and break-up in air-sparged bubble columns,"AIChE Journal, vol. 36, no. 10, pp. 1485–1499, 1990.

20. T. O. Oolman and H. W. Blanch, "Bubble coalescence in stagnant liquids," Chemical Engineering Communications, vol. 43, no. 4–6, pp. 237–261, 1986.

21. A. K. Chesters and G. Hofman, "Bubble coalescence in pure liquids," Applied Scientific Research, vol. 38, no. 1, pp. 353–361, 1982. ··

22. R. D. Kirkpatrick and M. J. Lockett, "The influence of approach velocity on bubble coalescence,"Chemical Engineering Science, vol. 29, no. 12, pp. 2363–2373, 1974.

23. J. W. Kim and W. K. Lee, "Coalescence behavior of two bubbles in stagnant liquids," Journal of Chemical Engineering of Japan, vol. 20, no. 5, pp. 448–453, 1987.

24. J. C. Rotta, Turbulente Stromungen, B. G. Teubner, Stuttgart, Germany, 1974.

25. H. Anglart and O. Nylund, "CFD application to prediction of void distribution in two-phase bubbly flows in rod bundles," Nuclear Engineering and Design, vol. 163, no. 1-2, pp. 81–98, 1996.

26. R. T. Lahey Jr. and D. A. Drew, "The analysis of two-phase flow and heat transfer using a multidimensional, four field, two-fluid model," Nuclear Engineering and Design, vol. 204, no. 1–3, pp. 29–44, 2001. ··

27. J. B. Joshi, "Computational flow modelling and design of bubble column reactors," Chemical Engineering Science, vol. 56, no. 21-22, pp. 5893–5933, 2001. ··

28. R. Clift, J. R. Grace, and M. E. Weber, Bubbles, Drops, and Particles, Academic Press, New York, NY, USA, 1978.

29. J. B. Joshi, U. V. Parasu, Ch. V. Prasad, et al., "Gas holdup structures in bubble column reactors,"Proceedings of the Indian National Science Academy, vol. 64, pp. 441–567, 1998.

30.   I. Zun, "The transverse migration of bubbles influenced by walls in vertical bubbly flow," International Journal of Multiphase Flow, vol. 6, no. 6, pp. 583–588, 1980.

31.   N. H. Thomas, T. R. Auton, K. Sene, and J. C. R. Hunt, "Entrapment and transport of bubbles by transient large eddies in turbulent shear flow," in Proceedings of the BHRA International Conference on the Physical Modelling of Multi-Phase Flow.

32.   D. A. Drew and S. L. Passman, Theory of Multicomponent Fluids, Springer, New York, NY, USA, 1999.

33.   A. Tomiyama, "Drag, lift and virtual mass forces acting on a single bubble," in Proceedings of the 3rd International Symposium on Two-Phase Flow Modelling and Experimentation, pp. 22–24, Pisa, Italy, September 2004.

34.   S. P. Antal, R. T. Lahey, and J. E. Flaherty, "Analysis of phase distribution in fully developed laminar bubbly two-phase flow," International Journal of Multiphase Flow, vol. 17, no. 5, pp. 635–652, 1991.··

35.   M. A. Lopez de Bertodano, Turbulent bubbly two-phase flow in a triangular duct, Ph. D. dissertation, Rensselaer Polytechnic Institute, 1992.

36.   M. Ishii and N. Zuber, "Drag coefficient and relative velocity in bubbly, droplet or particulate flows," AIChE Journal, vol. 25, no. 5, pp. 843–855, 1979.

37.   T. R. Auton, "The lift force on a spherical body in a rotational flow," Journal of Fluid Mechanics, vol. 183, pp. 199–218, 1987.

38.   M. A. Lopez de Bertodano, R. T. Lahey Jr., and O. C. Jones, "Phase distribution in bubbly two-phase flow in vertical ducts," International Journal of Multiphase Flow, vol. 20, no. 5, pp. 805–818, 1994.

39.   M. Lance and M. A. Lopez de Bertodano, "Phase distribution and wall effects in bubbly two-phase flows," Multiphase Science and Technology, vol. 8, no. 1–4, pp. 69–123, 1994.

40.   A. Tomiyama, A. Sou, I. Zun, N. Kanami, and T. Sakaguchi, "Effects of Eotvos number and dimensionless liquid volumetric flux on lateral motion of a bubble in a laminar duct flow," in Advances in Multiphase Flow, pp. 3–15, 1995.

41. A. Tomiyama, H. Tamai, I. Zun, and S. Hosokawa, "Transverse migration of single bubbles in simple shear flows," Chemical Engineering Science, vol. 57, no. 11, pp. 1849–1858, 2002. ··

42. Y. Sato and K. Sekoguchi, "Liquid velocity distribution in two-phase bubble flow," International Journal of Multiphase Flow, vol. 2, no. 1, pp. 79–95, 1975.

43. R. D. S. Cavalcanti, S. R. D. Neto, and E. O. Vilar, "A computational fluid dynamics study of hydrogen bubbles in an electrochemical reactor," Brazilian Archives of Biology and Technology, vol. 48, pp. 219–229, 2005.

44. J. Y. Tu, G. H. Yeoh, G. C. Park, and M. O. Kim, "On population balance approach for subcooled boiling flow prediction," Journal of Heat Transfer, vol. 127, no. 3, pp. 253–264, 2005. ··

45. D. Legendre and J. Magnaudet, "The lift force on a spherical bubble in a viscous linear shear flow,"Journal of Fluid Mechanics, vol. 368, pp. 81–126, 1998.

46. G. Kocamustafaogullari, W. D. Huang, and J. Razi, "Measurement and modeling of average void fraction, bubble size and interfacial area," Nuclear Engineering and Design, vol. 148, no. 2-3, pp. 437–453, 1994.

47. A. Iskandrani and G. Kojasoy, "Local void fraction and velocity field description in horizontal bubbly flow," Nuclear Engineering and Design, vol. 204, no. 1–3, pp. 117–128, 2001. ··

2

Predicting and Preventing Flow Accelerated Corrosion in Nuclear Power Plant

Bryan Poulson

School of Chemical Engineering and Advanced Materials, University of Newcastle, Benson Building, Newcastle upon Tyne NE1 7RU, UK

ABSTRACT

Flow accelerated corrosion (FAC) of carbon steels in water has been a concern in nuclear power production for over 40 years. Many theoretical models or empirical approaches have been developed to predict the possible occurrence, position, and rate of FAC. There are a number of parameters, which need to be incorporated into any model. Firstly there is a measure defining the hydrodynamic severity of the flow; this is usually the mass transfer rate. The development of roughness due to FAC and its effect on mass transfer need to be considered. Then most critically there is the derived or assumed functional relationship between the chosen hydrodynamic parameter and the rate of FAC. Environmental parameters that are required, at the relevant temperature and pH, are the solubility of magnetite and the diffusion coefficient of the relevant iron species. The chromium content of the steel is the most important material factor.

INTRODUCTION

Flow accelerated corrosion (FAC) of carbon steels in water has been a major concern in civil nuclear power production for over 40 years [1, 2]. The important features of FAC are the linear or increasing rate with time and the generation of a scalloped surface. Its effects have been unique in two important ways. Firstly it has affected nearly every reactor type worldwide and sometimes in more than one location; Figure 1 shows some examples [3–6]. Secondly it is probably the only corrosion mechanism that has led to accidents that have caused fatalities. There have been pipe ruptures leading to a release of steam and deaths of workers, but it must be emphasized that such fatalities are not unique to nuclear plants.

(a)

Plant	Component	Parameters				
	T (°C)					
		pH	O_2	d (mm)	Re	

a	Hinkley AGR[3]	SG tube inlet with orifce $d/d_0 = 3.28$ AVT secondary water	155	9.1-9.4	~ 2 ppb	15.6	2×10^5
b	Mihama PWR[4]	Condensate water pipe afer orifce $d/d_0 = 1.612$	140-142	8.6-9.3	<5 ppb	540	5.8×10^6
c	Surry PWR[5]	90° bend afer reducing T-piece in condensate system	190	8.9-9.0	4 ppb	305	10^7 ish
d	CANDU[6]	Bend afer end-ftting/outlet feeder pipe; primary water	305-315	10.2-10.8	~ 0	38-90	$3.5\text{-}7.7 \times 10^6$

(b)

Figure 1: Examples of FAC in nuclear power plants.

The occurrence of FAC, or erosion corrosion as it is sometimes known, is critically dependent on the following: the temperature and chemistry of the environment (pH and oxygen content), the hydrodynamics of the system, and the composition of the steel particularly the chromium content; this is shown schematically in Figure 2.

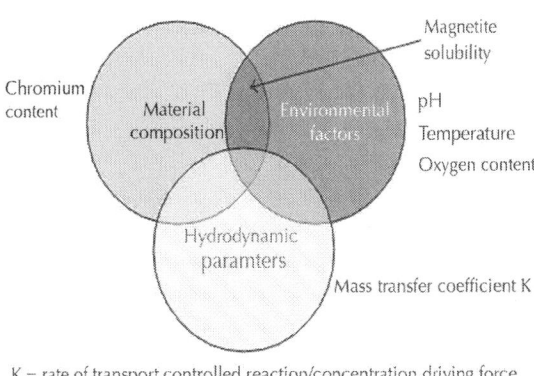

$$K = \text{rate of transport controlled reaction/concentration driving force}$$
$$Sh = aRe^xSc^y$$

Figure 2: Factors influencing FAC.

There are various approaches to predicting the possibility and the rate of flow accelerated corrosion. Testing has involved either actual components or a chosen specimen. In addition there are theoretical or empirical models, some computer based, available to allow the prediction of attack.

It is readily apparent that any review must be very selective; other reviews are available [7–10]. This review has tried to cover areas that the author has been involved with over the last 40 years that have been neglected or are contentious. It follows earlier reviews [11–14] but focuses on FAC of steels in the nuclear power plant industry. For both practical and mechanistic reasons there is a need to identify the hydrodynamic parameter which controls the occurrence and rate of erosion corrosion. It was previously suggested [12] that there was a spectrum of mechanisms which could be involved in erosion corrosion. This review concentrates on the corrosion end of the spectrum. It is argued that for dissolution based mechanisms the important hydrodynamic parameter is the mass transfer coefficient (K). The relationship between the rate of FAC and K is discussed in some detail and the development and effect of surface roughness on both K and FAC are considered. Finally some aspects of the solubility of magnetite and the prediction and prevention of FAC are discussed. This review is not a best buy guide to predictive programs.

Hydrodynamics

The various hydrodynamic parameters that have been credited with controlling the occurrence and rate of flow assisted corrosion are as follows.

- Velocity (V).
- Reynolds number (Re).
- Mass transfer coefficient (K).
- Surface shear stress (τ).
- Intensity of turbulence (TI).
- Freak energy density (FED).

An attempt to describe how each of these parameters might be measured is given in Table 1. It is widely accepted that it is neither the velocity nor the Reynolds number that is critical. Both the surface shear stress and the mass transfer coefficient are widely believed to be important. The former in the oil and gas field the latter in the power generating industry. But recently τ has gained some devotees [15] in the power generating industry. However there are some workers [16] who believe the following.

The mass transfer coefficient is intimately linked to wall shear stress, and the two cannot be practically separated either experimentally or mathematically [16].

Table 1: Possible important parameters and some measurement techniques

Parameter	Some measuring techniques	References
The velocity (V).	Calculated from flow rate and flow area Ultrasonic sensors Pitot tube	[11, 17, 18].
Reynolds number (Re).	Calculated from velocity, diameter of tube, and kinematic viscosity	[11, 17, 18].
Mass transfer coefficient (K). Requirements: Use of realistic geometries Measurement of both smooth and rough surfaces Use in two-phase flow	Dissolution of sparingly soluble solid Limiting current Analogy with heat transfer Computational	[12, 19–23].
The surface shear stress (τ) Requirements as for K	On isolated surface element using force transducer Pressure drop Various types of heat transfer sensors Limiting current density on isolated small electrode, but not for separated flows Computational	[12, 18–20, 24, 25].

Intensity of turbulence (TI) Requirements as for K	Analysis of fluctuations in limiting current density on isolated small electrode. Hot wire Laser-Doppler anemometer	[11, 18–20].
Freak energy density (FED) Requirements as for K	Complex analysis of fluctuations in limiting current density on isolated small electrode, single author use.	[26–28].

This view has been challenged by a number of workers [25, 29] based on the breakdown of the analogy between mass, heat, and momentum in detached flow. The consequences of this are that in detached flow the surface shear stress cannot be calculated from the measured limiting current density (LCD) at an isolated microelectrode, where in normal flow τ is proportional to the cube of the limiting current density (Leveque equation); this is demonstrated in Figure 3. This problem seems to have been ignored by Schmitt and Gudde [30] whose approach was to obtain a functional relationship between shear stress and limiting current density for a normal flow situation (Figure 4). Then to use the same relationship to convert LCDs measured in disturbed flow to shear stresses (Figure 4). Apart from being invalid this approach involved the extrapolation of a maximum shear stress in the channel wall of under $60\,N/m^2$ to over $14000\,N/m^2$ in the disturbed flow.

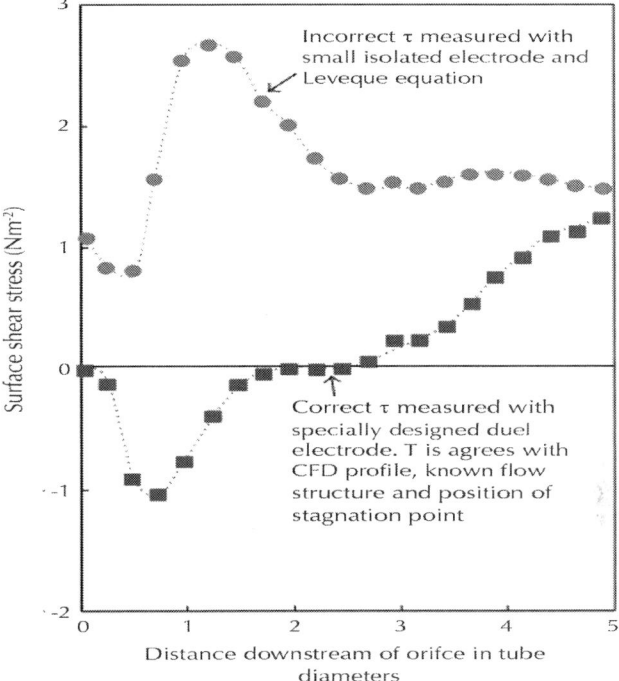

Figure 3: Correct and incorrect electrochemical measurements of surface shear stress, after [25].

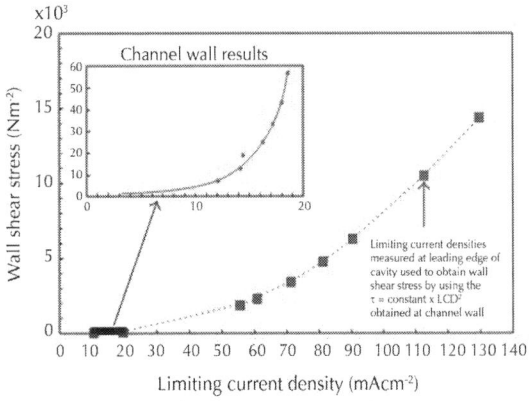

Figure 4: Schmitt's measurement of surface shear stress in detached flow (after [30]).

One way of refuting the importance of τ is its variation, for example, downstream of an orifice where the FAC and shear stress profiles are fundamentally different (see Figures 3 and 5). This refutation of the importance of τ has apparently been misunderstood [32]. A similar rejection of the dominant role of τ has been made by Matsumura et al. [33, 34] in work with an impinging jet. Another reason to believe that τ is unimportant is because it seems that the surface shear stress is not sufficient to produce the effects ascribed to it, namely, mechanically disrupt a surface film.

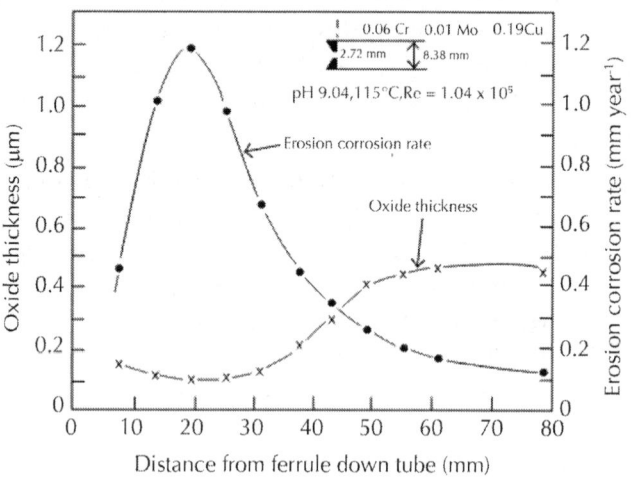

Figure 5: Correlation of FAC rate and oxide thickness ([31]).

For copper alloys in seawater Matsumura showed that the FAC profile did not correspond to the measured shear stress distribution on an impinging jet specimen but could be explained by the forces, measured with a pressure transducer, due to the peak in turbulence intensity at about 2 jet diameters from the centre of the specimen. Of course mass transfer is caused by bulk convection and turbulent convection, so while ruling out the primacy of the importance of shear stress this does not prove the importance of mechanical factors. It must be remembered that increased mass transfer can dissolve away a protective surface layer; this was elegantly analysed by Coney [35].

Schmitt and Mueller initially proposed [36] a fatigue based model for oxide failure, and it is not clear to this author why such an approach was apparently rejected for the freak energy density (FED) model. Although this was applied to the oil and gas industry [27] his ideas have been published in power plant chemistry [28] and thus warrant discussion. The FED model is conceptually similar to Matsumura's view on the importance of the turbulent flow generating high local stresses.

It was found that forces in high energy microturbulence elements oriented perpendicularly to the wall are finally responsible for the scale destruction (freak energy density (FED) Model). The maximum FED in a flow system can be measured with appropriate tools; however, in practical cases this is not necessary, because in a given flow system the FED is proportional to the wall shear stress by a factor in the order of 10^5 to 10^6. Thus, for practical application and flow system evaluations wall shear stresses can be used to quantify flow intensities in given flow systems [27].

FED is a derived freak value from measuring the noise in the limiting current density (LCD) on isolated small point electrodes (ISPE). It appears [27, 28] that this is a complex process involving wavelet transforms, simulations, and phasing in of waves. The net result is a freak current density; values of as high as $500\,A\,cm^{-2}$ have been quoted. The energy density of a freak wave volume element accelerated towards the surface can be expressed according to classic wave dynamics from

$$w = \frac{dE}{dV} = 0.5\rho A^2 \omega^2,$$

(1)

Where w is energy density (Pa), ρ is density (kg/m^3), is wave amplitude (m), and ω is wave frequency (s^{-1}). Both energy density and wall shear stress have the unit Pascal in common. It was, therefore, assumed that the same relation used in the Leveque equation can be used to calculate the freak energy density from the freak current density. For the freak current density of $500Acm^{-2}$ a freak energy density of $3\,GPa$ was derived.

Such an approach can be questioned on a number of important points.

- Although there are time response limitations involved in LCD measurements it might be expected that some freak events could at least be partially detected, after all real freak waves can be observed.

- No attempt to measure such high stresses was made, for example, by fast response pressure transducers, as Matsumura did.

- The use of the Leveque equation to obtain freak energy densities from freak current densities is highly questionable.

- There is not a single relationship between τ and FED [27, 28] and from a practical point of view this would have to be obtained for each geometry of interest. And measuring surface shear stresses in detached flow is, as outlined earlier, not possible using Schmitt's technique.

- How rough surfaces that develop naturally during corrosion are dealt with is unclear.

Also as yet there has been no data that supports the use of FED to predict FAC, indeed Figure 6 from a 2009 paper [27] appears key since it suggests that there is not a single relationship between corrosion rate and FED, for two different flow conditions, in the same environment. Finally there is a mechanistic problem in what exactly happens when a surface layer is cracked; Schmitt has suggested the following.

Once removed the high local flow intensities prevent the re-formation of protective layers and, hence, start fast mass transport controlled local metal dissolution (FILC, also called erosion corrosion).

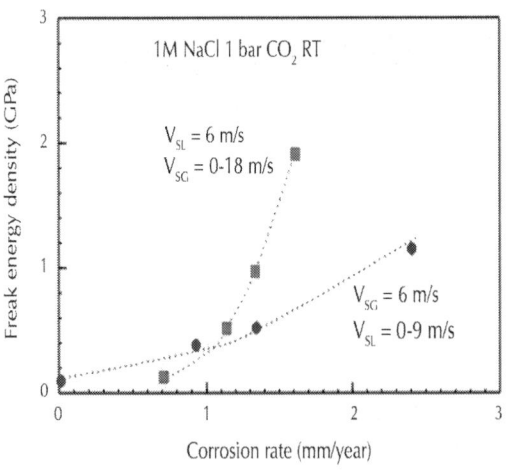

Figure 6: No single relationship between freak energy density and FAC from specimens exposed to gas pulsed impinging jet (after [27]).

For steels in pure boiler water there is good evidence that the protective magnetite layer is thinned (not cracked) and the FAC rate correlates with this thinning [31] and the mass transfer coefficient (Figure 5). There is also the possibility that the mass transfer could be large enough to lead to the film being completely removed by dissolution. In this case the subsequent rate of attack would probably be limited by activation kinetics. This film dissolution appears to have occurred on carbon steel exposed to sodium nitrate solutions under an impinging jet [11].

Of course in most situations in which corrosion is occurring the surface undergoes some roughening; this then raises the question of how to measure the changes in the hydrodynamic parameter as the roughness develops. Roughness development and other considerations (Table 1) led the current author to develop, justify, and apply the dissolution of copper in ferric chloride solutions [21, 37–41] (Figure 7), to the measurement of mass transfer coefficients; typical specimens from single-phase and two-phase studies are shown in Figures 8 and 9.

Anodic reaction: Cathodic reaction: Transport of Fe^{3+} to the surface is rate controlling

$Cu \quad CuCl_2^- \text{ or } CuCl_3^{2-} \quad Fe^{3+} \quad Fe^{2+} + e^-$

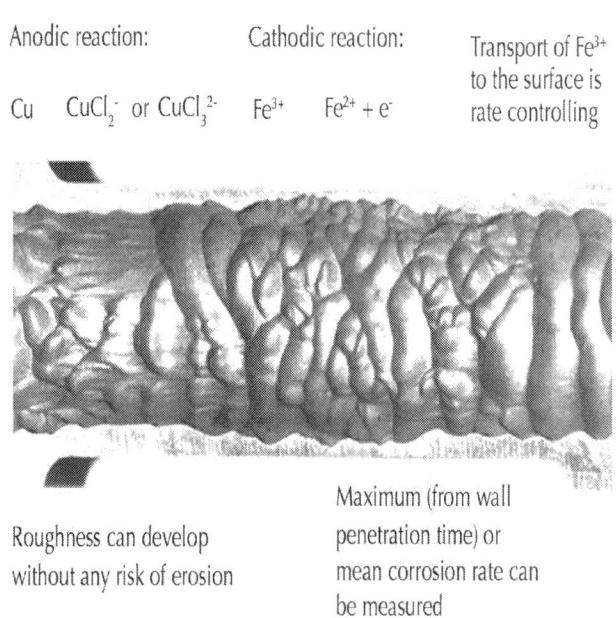

Roughness can develop without any risk of erosion

Maximum (from wall penetration time) or mean corrosion rate can be measured

Figure 7: Copper specimen used to measure K in acid ferric containing solutions ([21,37–41]).

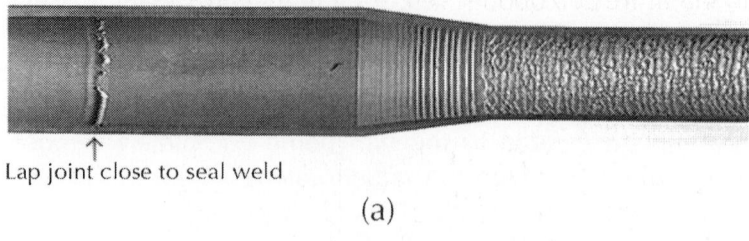

Lap joint close to seal weld

(a)

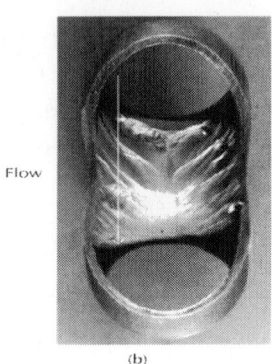

Flow

(b)

Re ~ 4 × 10⁴

Re ~ 3 × 10⁵

(c)

Figure 8: Typical copper specimens after testing in single phase flow. (a) Reducer (39 to 25.6 mm), Re 2.8 × 10⁵ in larger tube ([40]). (b) 180°,

2.5 D bend at Re of 70000 ([37]). (c) Impinging specimens at 9.5 mm from 9.5 mm diameter jet ([40]).

(a)

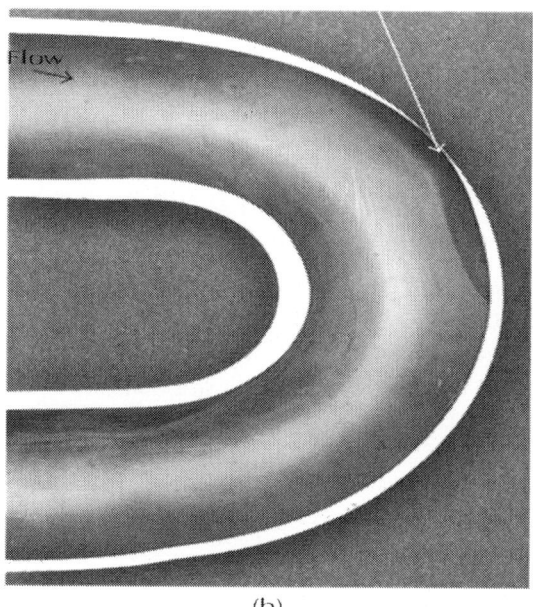

(b)

Figure 9: Typical copper specimens after testing in two-phase flow. (a) 10-bar submerged gas jet at 1 mm from specimen ([13]). (b) Annular two-phase flow ([39]).

Because some predictive models use an enhancement factor to describe a geometries effect relative to a straight tube such a factor is often quoted. It must be emphasised that this factor is usually a function of the Reynolds number and its use has nothing to do with—the mass transfer coefficient at the point of interest becomes difficult to measure or even define, so "enhancement factors" over the straight-pipe values are employed [32].

Recently there has been an increasing tendency to use computation fluid dynamics (CFD) techniques to obtain hydrodynamic parameters for subsequent use in FAC assessments. CFD has the ability to investigate very high Re's, which often cannot be reached in experimental studies. However CFD results must be shown to agree with well established data before it can be applied to the higher flows. A number of applications of CFD are listed in Table 2; unfortunately some of these computations have been carried out without any comparison to well established experimental data; in some cases there are significant differences. This is particularly evident after the Mihama-3 FAC failure downstream of an orifice; the mass transfer characteristics of which have been studied by a number of workers and reviewed by others. Figures 10 and11 are a summary of and comparison with excepted correlations. The difference between the two CFD predictions is of some interest; the higher results may be a misprint in units, as the other data agrees very well with a simple calculation for fully developed flow. It is of great interest that the CFD predicted enhancements are much higher than those from expected correlations downstream of an orifice, yet no comparison with the well established lower Reynolds number data were made. Also in this authors' view CFD has not been shown to be capable of modelling the roughness and its effects that develops naturally by metal dissolution or oxide deposition.

Table 2: Examples of the use of CFD in FAC

Authors	Purpose	Comments	References
Uchida et al.	Prediction of FAC with Mihama as test case	No comparison with existing data and its extrapolation to high Re's or effects of roughness	[42]
Hoashi et al.	Prediction of FAC with Mihama as test case	No comparison with existing data and its extrapolation to high Re's or effects of roughness.	[43]

Pietralik and Smith, Pietralik and Schefski	Prediction/explaining CANDU feeder FAC	Compares with some but not all bend data. Attempts to deal with roughness development and component interactions	[44, 45]
Nesic and Postlethwaite	Predict e-c following sudden expansion	Makes key point that profiles of K and shear stress do not correlate	[29]
Zinemams and Herszaz	Predict FAC in bifurcation and nozzle	Limited detail to check results but claims there is agreement, effects of roughness not considered.	[46]
Yoneda	Predict FAC in PWR and BWR, of 45° elbow	Interesting but again no comparison with others or the effects of roughness	[47]

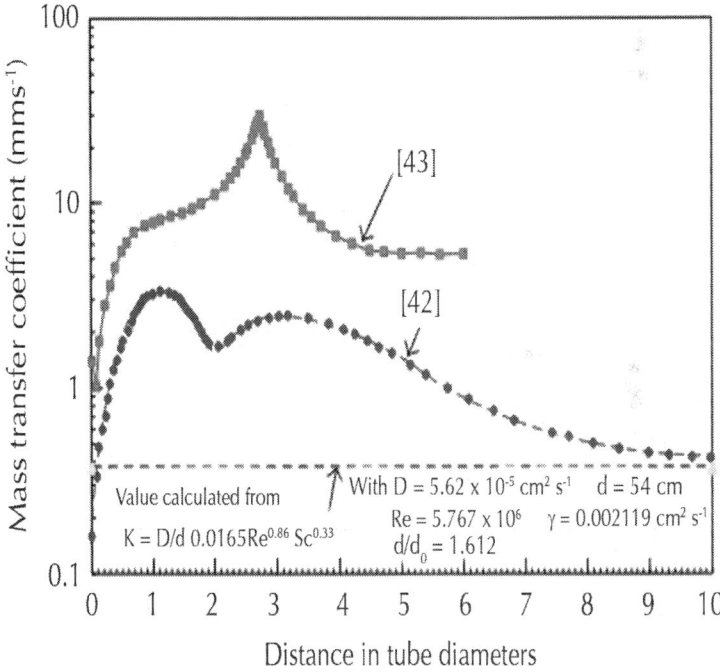

Figure 10: Comparison of mass transfer profile for Mihama.

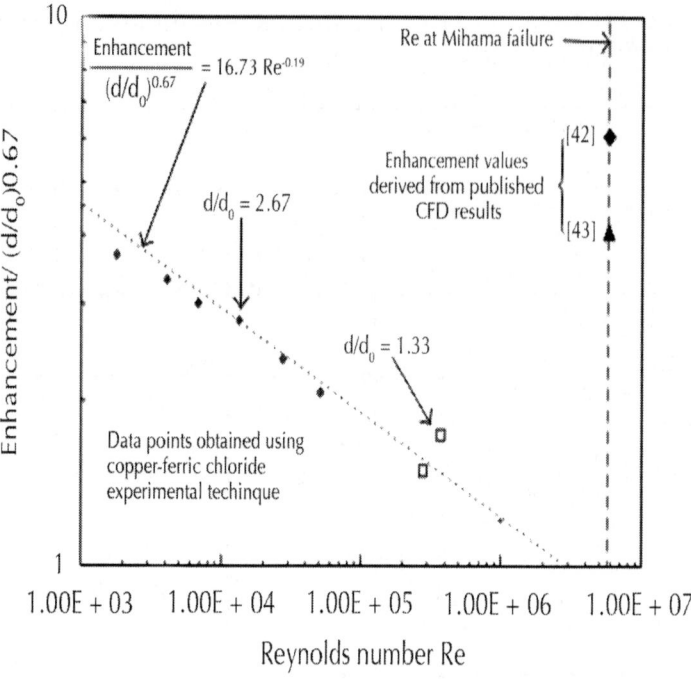

Figure 11: Comparison of accepted mass transfer correlation and results from Cu/Fe³experiments and CFD values for Mihama.

Roughness development in situations where a solid is dissolving under mass transfer control has been investigated by a number of workers. One theory attributes roughness to the imprinting of the fluid on the surface [48] and the other to the role of surface defects [49]. Criteria for the development of roughness have not been fully developed and it is perhaps overlooked that there is a tendency for leveling of any surface which is dissolving under diffusional control [39]. For example, consider an isolated roughness element exposed to single phase flow. It would be expected that the rates of dissolution of both the peak of the roughness element and the region after the roughness element were both higher than the region upstream. For roughness to develop the region after the roughness peak must dissolve faster than the roughness peak itself. In two phase flow it is probable that the annular film would leave the wall after the peak rather than inducing separated flow as in single phase conditions. Thus it would seem reasonable that in annular two-phase flow roughness development would be more difficult and would tend to occur only when the thickness of the wall film was such that a recirculation zone could form.

It is the effect of roughness that is of general relevance, and our findings [37–41] can be summarized as follows.

- While defects can produce surface roughness, they are not a requirement. The roughness that develops usually reflects the flow structure. For example, Figure 7 illustrates the symmetrical peaks at the inlet to the 180° bend and evidence of counter rotating vortices in the bend region.

- There appears to be a critical Re required for roughness to develop. The critical Re value ~5 × 10^4 in 8 mm diameter tubing [40] is significantly higher than that found using dissolving plaster of Paris [22], where roughness developed at the lowest Re tested of 1.9 × 10^4 in 25 mm diameter tubing.

- There are other indications [39] that using the dissolving plaster of Paris technique can give misleadingly high values of K due to erosion of the plaster occurring.

- If the surface roughens the mass transfer can increase and there are indications [40] that the roughness becomes more important then than the geometry in influencing mass transfer.

- An upper bound mass transfer correlation for all rough surfaces has been proposed [38] as shown in Figure 12:

$$Sh = 0.01 \ Re \ Sc^{0.33}. \qquad (2)$$

- There is some evidence, Table 3 [41], that the development of roughness has a smaller effect on geometries where flow is separated, as compared to normal flow. In the former turbulence is generated away from the wall, while in normal flow it is generated close to the wall. This was first realized with the region downstream of an orifice in that some enhancement in mass transfer occurred in some tests but again nowhere near a Re_0 dependency (Figure 10). Also for a multi-impinging jet and a cylinder in restricted cross flow the rough surface mass transfer correlations all have a Reynolds number dependency with an exponent less than one, Figure 13, with

$$Sh = 0.195Re^{0.65}Sc^{0.33} \quad \text{Multi-impinging jets,}$$

$$Sh = 0.019Re^{0.87}Sc^{0.33} \quad \text{Cylinder in restricted cross flow.} \qquad (3)$$

- If roughness develops, the height (ε) and wavelength (λ) of the roughness elements decrease with increasing Re [40]. This means that the enhancement in mass transfer caused by roughness does not increase with increases in ε/d. There is a suggestion [50] that the roughness that develops at any Re produces the maximum resistance to flow..

- If roughness develops the enhancement factor of any geometry over a straight tube is not a constant [37] and will usually increase with the Reynolds number. For the region downstream of an orifice the enhancement decreases with increases in Re for both smooth and rough surfaces.

In summary the initial surface roughness (or defects) will influence the subsequent development of dissolution induced roughness which will occur at lower flow rates for rougher initial surfaces. However the final or equilibrium roughness that develops as a result dissolution will probably be largely dependent on the flow rate, with smaller scallops as the Reynolds number increases. The effect of roughness developing, in increasing the rate of mass transfer and thus the rate of FAC, is expected to be less for geometries where flow separation or detachment occurs.

Table 3: Differences between normal and separated flow

	Normal flow	**Separated flow**
Shear stress and K	Related	Not related
Turbulence created	Near wall	Away from wall
Roughness effects on K and ΔP	Both increase	Evidence for increase lacking
Roughness effects on FAC	Limited clear evidence of increase	Evidence for increase lacking

But in all probability there will be a number of situations that are between these two extremes.

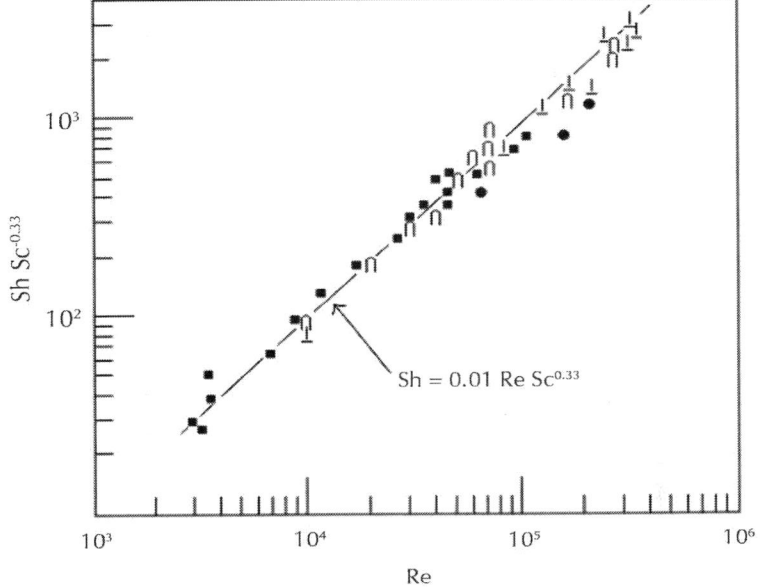

(a)

Symbol	Geometry	Technique	Ref
∩	2.5d 180 bend in 22.6 mm tube	Copper corrosion	Poulson and Robinson 1988
⊥	9.5 mm d impinging jet at height of $1d$		
	Rotating cylinder $\frac{d}{\varepsilon}$ =87	LCDT	Kapperesser et al. 1971
•	90mm pipe	Plaster dissolution	Wilkin and Oates 1985

(b)

Figure 12: Poulson's rough surface correlation ([38]).

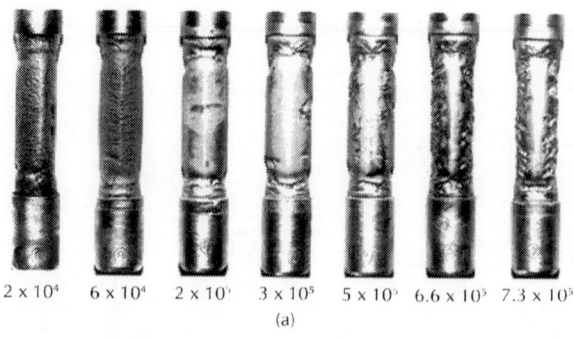

2×10^4 6×10^4 2×10^5 3×10^5 5×10^5 6.6×10^5 7.3×10^5

(a)

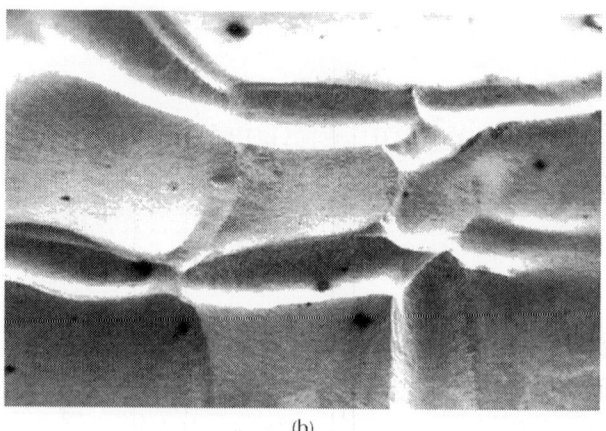

(b)

Figure 13: Examples of specimens after testing under detached flow. (a) Cylinders in cross flow at stated Re's: $Sh_{max} = 0.019Re^{0.87}Sc^{0.33}$. (b)Multi-impinging jet (d = 1.5mmat distance of 2 mm): $Sh_{average} = 0.195Re^{0.65}Sc^{0.33}$.

In two-phase flow it is not immediately obvious how the mass transfer data can be simply incorporated since there is some evidence [39] that Chen's correlation is not valid. Figure 14 shows a simple model can apparently be used to relate the tightness of bends to the resulting influence of the annular flow on the mass transfer at bends. There is evidence that mass transfer effects dominate at least up to 50ms^{-1}; but clearly mechanical damage will occur above some critical velocity which needs to be defined; this leads to the concept of a spectrum of mechanisms as suggested earlier [12]. Models ascribing droplet impingement causing mechanical damage are referenced later in predictive models but are outside the scope of this paper.

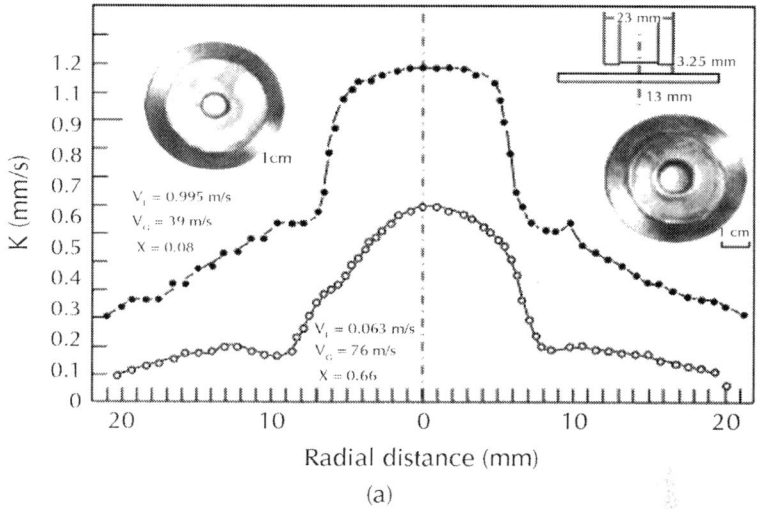

(a)

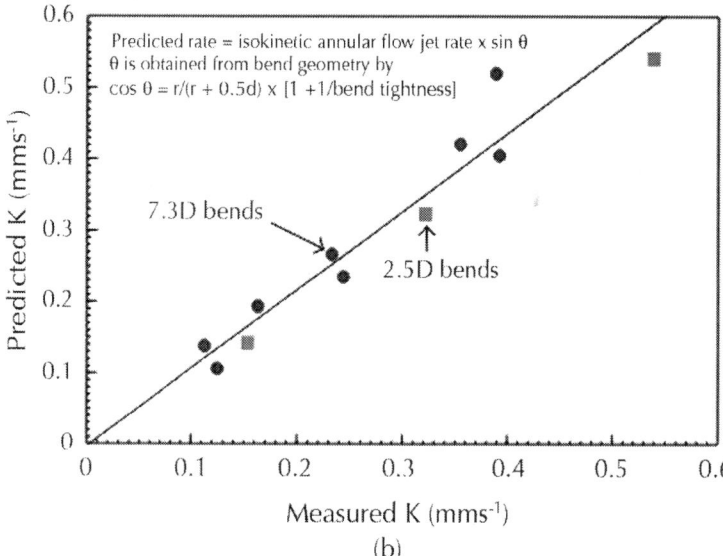

(b)

Figure 14: (a) Isokinetic annular flow impinging jet specimens ([39]). (b) Prediction of mass transfer at bends from jet data ([39]).

The relationship between K and FAC must not be assumed to be linear; this is discussed later in this review.

ENVIRONMENTAL VARIABLES

It is widely accepted as shown schematically in Figure 1 that the three most important environmental parameters influencing the occurrence and rate of FAC are as follows.

- Temperature.
- pH.
- Oxygen content.

These have been discussed in detail and further discussion is beyond the scope of this review. However each of these variables probably exerts their influence through their effect on the solubility of magnetite. It is believed that this has been a neglected topic. For example in one of the first reported cases of FAC at 300°C it was suggested [51] to be the result of high mass transfer and low Cr content in the steel with little consideration to the reduced solubility of the magnetite at that temperature. A similar situation occurred in a recent paper [52] describing low temperature FAC.

In general solubility's can be calculated from available thermodynamic data or measured. For a very sparingly soluble oxide like magnetite the preferred option has been to experimentally determine it using equipment such as that shown in Figure 15. There are a number of experimental difficulties such as the need to establish steady state conditions under controlled redox conditions, the problems with colloidal material, and the influence of trace impurities. Such solubility data, for example, Figure 16, is then used to obtain thermodynamic data for the various possible dissolving species, for example, Figure 17. The total solubility of iron species can then be modeled and predictions can be made (Figure 18).

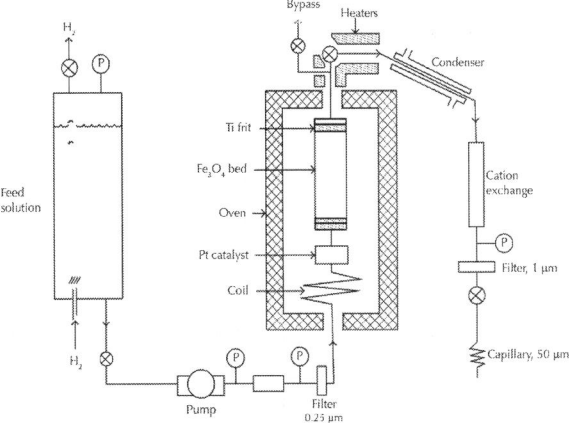

Figure 15: Typical equipment for solubility measurements (after [53]) and possible problems. Typical potential problems might include (1) fine particles passing through filters, (2) fine grain size and surface energy effects, (3) purity of materials, and (4) corrosion of equipment influencing results. (5) Bignold ([54]) suggested that magnetite is a mixture of ferrous and ferric ions. Ferric ion solubility is very low and on isolated magnetite ferric ions will build up on surface. Is it possible that this could lead to apparently low solubility and the true solubility should be obtained from magnetite coated iron?

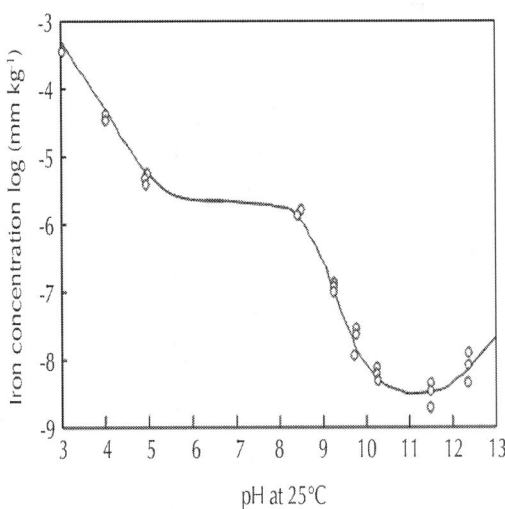

Figure 16: Typical results of solubility measurements (after [53]).

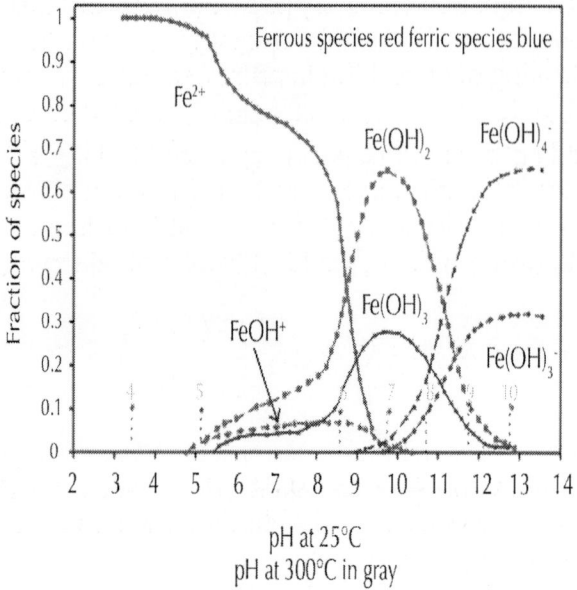

Figure 17: Defining dissolving species (after [53]).

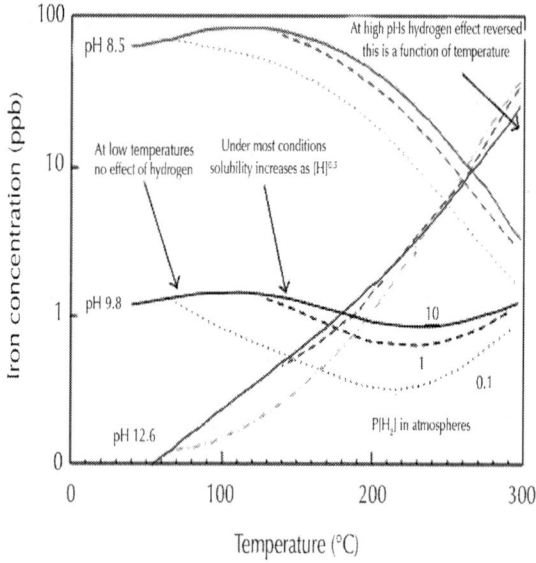

Figure 18: Examples of model predictions (after [55]).

From an examination of the key papers [53, 55–58] there are certain key facts, some are well known, for example, 1 and 2, and form the basis of controlling of deposition in primary circuits, others are not, yet they could significantly influence the ability to predict FAC.

• The gradient of solubility with respect to pH changes from negative to positive at a critical pH, producing a typical U-shaped curve, Figure 16. The pH value this occurs at is of intense interest; an estimate (using T&LeB data [53]) can be obtained from

$$pH = 6.456 + 16.365 \exp\left(\frac{-T°C}{88.518}\right). \qquad (4)$$

• The gradient of solubility with respect to increasing temperature changes from negative at lower pH to positive at higher pH, Figure 18; this occurs at ~9.4 and 9.9 at temperatures of 300°C and 150°C, respectively [53].

• There are various functional relationships between iron solubility and hydrogen content which arise because of the details of the stable species and the nature of the oxide dissolution reaction, Table 4 and Figure 18.

• Usually [53, 56] the solubility is proportional to $[H_2]^{1/3}$:

$$\frac{1}{3}Fe_3O_4 + (2 - |b) H^+ + \frac{1}{3}H_2$$

$$\longleftrightarrow Fe(OH)_b^{2-b} + \left(\frac{4}{3} - b\right) H_2O \qquad (5)$$

• As pHs increases the solubility is predicted [53] to become proportional to $[H_2]^{-1/6}$,, and that will occur above a certain temperature [55]. Consider

$$\frac{1}{3}Fe_3O_4 + (3 - b) H^+$$

$$\longleftrightarrow Fe(OH)_b^{3-b} + \frac{1}{6}H_2 + \left(\frac{4}{3} - b\right) H_2O \qquad (6)$$

However the data appears to be insufficient to make accurate predictions of pH's and temperature's and the expectation is that the transition will be gradual.

- Solubility is independent of [H_2] at low temperatures below ~83°C [55]. But this temperature increases with increases in [H_2]. This is due to the activity of ferrous ions in solution being controlled by a hydrous ferrous oxide phase rather than magnetite [55].

- There appear to be significant differences between the two most quoted data sets as shown in Figure19. In particular the values of Tremaine and LeBlanc (T&LeB) are lower than those of Sweeton and Baes (S&B) and mirror the temperature dependency of FAC more closely.

- Under conditions where ferrous ions are dominant the effect of potential [59], Figure 20, or oxygen content of the water is profound and of great practical importance; the solubility of haematite is exceedingly small.

- There is no solubility data available for magnetite containing chromium; such data might allow the role of chromium to be clarified.There are different functional relationships between pH_{RT} and pH_{HT} for ammonia, morpholine, ethylamine, and lithium hydroxide, which depend on basicity, volatility, temperature, and steam quality which need to be considered; this has been discussed recently by Bignold [60].

Table 4: Possible variations in magnetite solubility dependency on hydrogen concentration

Regiem	Hydrogen dependency	Reference
Normally	$[H]^{1/3}$	[53, 55, 56]
At high pHs above ~220°C	$[H]^{-1/6}$	[53, 55]
At temperatures below ~80–110°C	Independent	[55]
At high temperatures above pH~9.8	Less than $[H]^{1/3}$	[55]

Such effects could influence both dissolution and deposition.

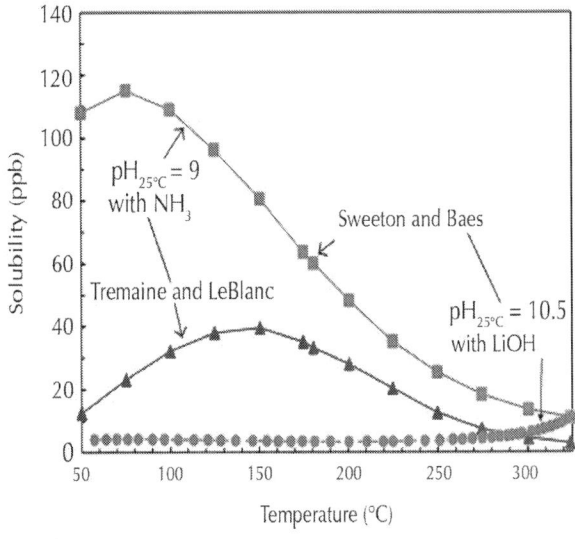

Figure 19: Comparison of solubility data ([2] with additions).

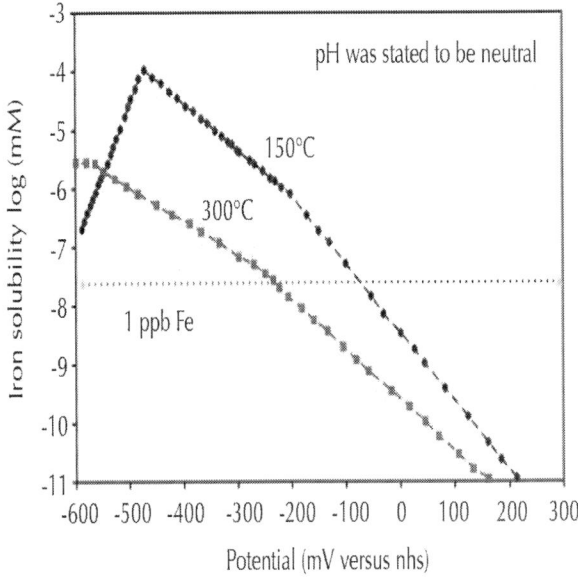

Figure 20: Effect of potential on solubility ([59]).

As indicated earlier another important parameter influenced by the environment is the diffusion coefficient (D) of the relevant dissolving iron species. This is rarely discussed, either the value utilized or how it was obtained. One of the few relevant discussions was presented by Coney [61]; it was indicated that there were a number of possible dissolving ions, and diffusion coefficients were calculated using both anionic and cationic data. As Newman [62] and others [12] have pointed out it is the ionic diffusion coefficient (cm²/s) which should be calculated using the Nernst-Einstein equation:

$$D_i = \frac{R\lambda_i T}{n_i F},$$ (6)

Where R the universal gas constant is 8.3143Jmoledeg, λ means the ionic equivalent conductance is in ohm-cm²/equiv, T is the temperature in degrees Kelvin, n is the charge on the ion, and F is Faraday's constant 96,487 C/equiv. The effects of temperature are usually calculated using Stokes-Einstein equation or Wilkes rule ($D\mu/T$ = constant) and this has been compared to an activation energy approach and found in reasonable agreement [12].

In Figure 21 the results of Coney are shown and compared to the recalculated values for FeOH⁺, which Coney suggests is the most relevant species. It appears that the recalculated values are approximately half of the earlier values and a correlating equation obtained using *Table CurVe*⁎ is given. Temperature also influences the density and viscosity of water; such values are readily available in the literature.

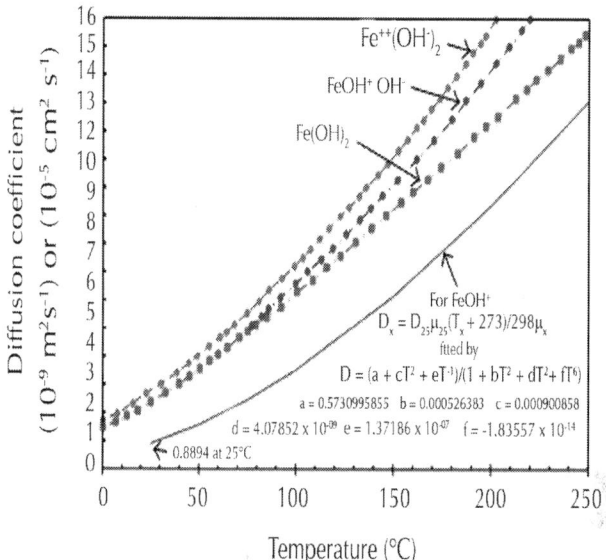

Figure 21: Examples of calculated diffusion coefficients; molecular data from [22, 61]: ionic data-current work.

Temperature also influences the density and viscosity of water; such values are readily available in the literature.

MATERIAL INFLUENCES

It has been known for some considerable time that the chromium content of steel [63, 64] has an important role in influencing flow accelerated corrosion; typical results are shown in Figures 22 and 23. There is some evidence [63] that under some situations copper and molybdenum may also be beneficial. Suggested correlations for the fractional reduction (FR) in FAC caused by alloying include

$$FR_{FAC} = \left(83Cr^{0.89}Cu^{0.25}Mo^{0.2}\right)^{-1},$$

$$FR_{FAC} = (0.61 + 2.54Cr + 1.64Cu + 0.3Mo)^{-1}. \qquad (8)$$

It is of considerable interest that welds can have weld metal lower or higher in chromium than the parent metal; this needs to be considered in any assessment.

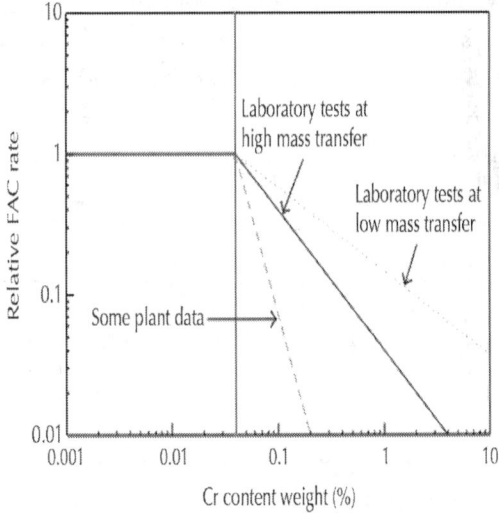

Figure 22: Effect of steels Cr content on FAC ([65] with modifications).

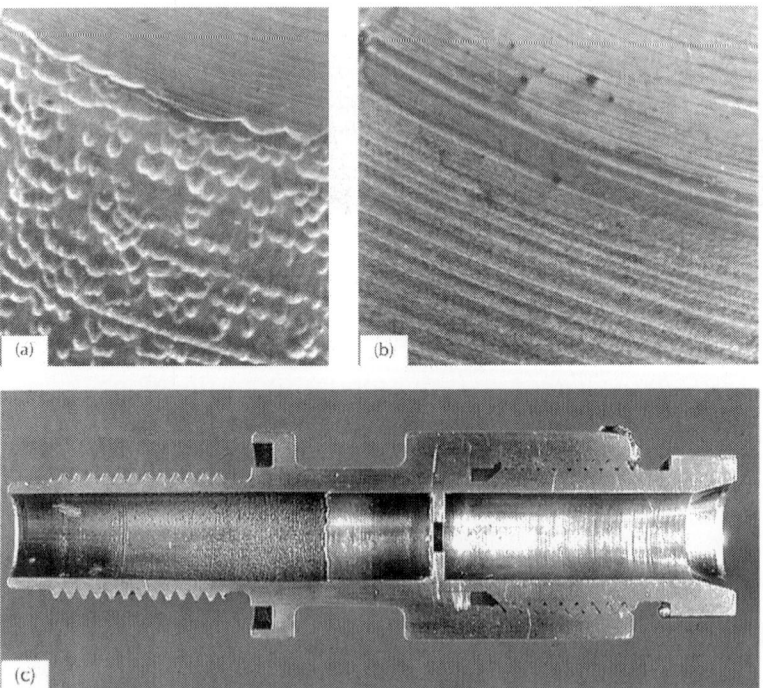

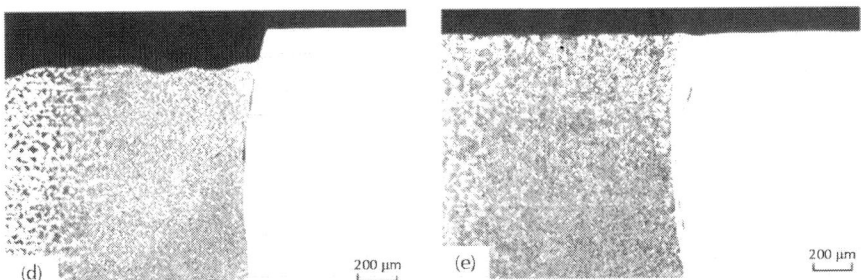

Figure 23: Effect of 1%Cr on FAC specimens after 2280 hours at 155°C at pH 9.1. (a) Surface of carbon steel. (b) Surface of 1Cr0.5 Mo steel. (c) Section of orifice holder. (d) Section through carbon steel-stainless steels. (e) 1Cr0.5Mo-stainless steels.

In general water chemistry changes are so much easier to implement than material changes; it is only in new or replacement components that material changes can be made.

It is believed that the first instance when the possibility of FAC influenced, indeed determined, the choice of material (1%Cr1/2%Mo) was for the economizers for the British advanced gas cooled reactors (AGRs) once through boilers at Torness and Heysham 2. This was substantiated by long term tests (22800 hour) in laboratory rigs, on-site rigs, and in situ SGs exposure. Both tight 180° bends and orifice geometries were utilized, Figure 24. Interestingly the feedwater system for the British PWR at Sizewell, ordered later, was specified as carbon steel. While there is no dispute as too the benefits of chromium additions there have been at least three suggestions as to its mechanism.

Figure 24: High pressure and high temperature flow loop for FAC studies on British AGR components.

- It slowly enriches at the dissolving magnetite surface and limits the solubility [66, 67].
- It alters the kinetics of oxide dissolution, possibly after enriching at the surface [68].
- It changes the oxide porosity [69].

PREDICTIVE APPROACHES

Testing

There are two possible approaches. The first is to test all geometric features of interest under realistic and accelerated conditions to obtain margins of safety. Figure 24 shows such a rig used to test AGR components at design and twice design velocities for times up to 22800 hours.

The alternative approach is to use a specimen with a range of mass transfer rates as illustrated in Figure 25, or a single specimen over a range of flow rates. In all cases it is good practice to check any relationship, between the suspected hydrodynamic parameter and the rate of FAC, with tests using a different specimen geometry.

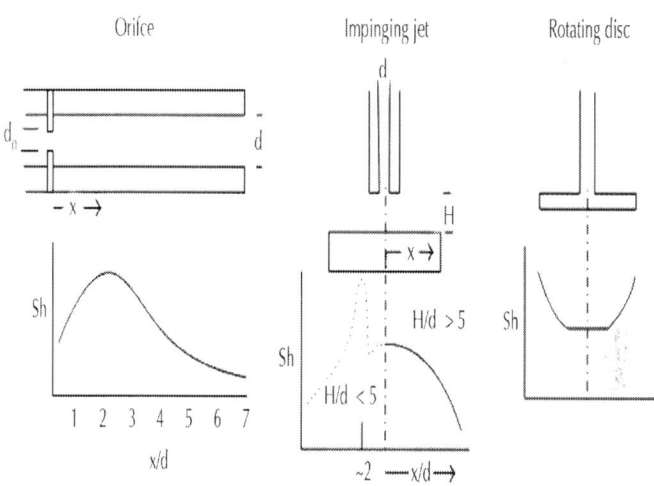

Figure 25: Specimens with mass transfer gradients ([12]).

The author can see no point in using a specimen for which there is no mass transfer data available. It is important in all cases to ensure the following.

- Tests are carried out long enough to obtain realistic steady state rates.

- The method of measuring the corrosion rate has the required precision; various methods have been suggested [12].

- The environmental conditions in the test rig really do correspond to the practical situation. The use of a marker geometry, that is, one which corrodes at a known rate, has much to commend itself. The advantage of the first approach is that real components can be tested and the dependency between corrosion rate and mass transfer will be established if the latter is known. It must be emphasized that this might change if the environmental conditions change. The advantage of the second approach is that it might prove possible to obtain the FAC versus K relationship with a single specimen where high precision measurements are possible. One difficulty in using specimens is that the development of roughness can be significantly different between geometries. The specific example of using small diameter tubes to obtain data relevant to large diameter pipes has been covered in some detail [40].

Theoretical Based Models

There appears to be reasonably widespread agreement that the basic mechanism of FAC is the enhanced transport of dissolved ferrous ions away from the metal surface. This causes the protective magnetite film to be thinned down and results in essentially linear kinetics. This is to be compared to parabolic kinetics during normal corrosion.

The simplest formulation of an FAC model gives the corrosion rate as the product of the mass transfer coefficient (K) and the solubility driving force (ΔC), where ΔC is difference between the solubility and the bulk solution level:

$$\mathrm{FAC}_{\mathrm{rate}} = K\Delta C; \tag{9}$$

K is usually obtained from correlations of the general form

$$\mathrm{Sh} = \mathrm{constant}\ \mathrm{Re}^x \mathrm{Sc}^y, \tag{10}$$

where Sh (the dimensionless mass transfer Sherwood number) is given by $K(d/D)$..

Re (the dimensionless Reynolds number) is given by (d/γ) and Sc (the dimensionless Schmitt number) is given by γ/D.

V is the relevant velocity (ms^{-1}), d is the characteristic specimen length (m), γ is the kinematic viscosity (m^2 s^{-1}), and D is the diffusivity of the relevant species (m^2 s^{-1}). Substituting for K,

$$\mathrm{FAC}_{\mathrm{rate}} = \mathrm{constant}\ V^x d^{x-1} \gamma^{y-x} D^{1-y} \Delta C. \tag{11}$$

So in principal, if the mass transfer is known and the concentration driving force is known, the only other parameters required are the diffusion coefficient of the relevant ionic species and the appropriate kinematic viscosity.

However, most if not all the predictive models involve additional processes shown in Figure 25. The differences in the various mechanistic models arise in the way the processes are quantified and combined, in particular in the explanation of the key environmental parameters especially the temperature and how the flow dependency of the measured FAC rates can be explained.

From a predictive aspect a very important difference is that some theories predict a linear relationship between mass transfer and flow accelerated corrosion rate: a nonlinear rate is predicted by other models particularly the CEGB theory of Bignold, Garbett, and Woolsey. Other workers have argued that a greater-than-linear dependency of rate on mass transfer equates to a mechanical contribution probably in removing oxide.

The CEGB model [10, 54, 66, 67, 70, 71] (Bignold, Garbett, and Woolsey) was developed because the maximum FAC rate and the measured FAC profile downstream of an orifice could not be explained by the available magnetite solubility data, and a linear relationship between FAC and mass transfer. In addition to explain the decrease of FAC at lower temperatures, it was postulated that the rate of magnetite dissolution (R of Fe_3O_4 dissolution) controls the rate of FAC at low temperatures:

$$\frac{1}{FAC_{rate}} = \frac{1}{K\Delta C} + \frac{1}{R \text{ of } Fe_3O_4 \text{ dissolution}}.$$

(12)

There is strong evidence that the solubility of magnetite is a function of the hydrogen content of the water. Thus it appears that the process could be self-accelerating in that the higher the corrosion rate is, the more hydrogen is produced locally, and this could enhance the solubility of the magnetite. An electrochemical model was developed which indicated that the solubility of magnetite could be proportional to the square of the mass transfer coefficient.

Thus the FAC rate could be proportional to the cube of the mass transfer coefficient, or in general with n being a variable between 1 and 3:

$$FAC_{rate} = K^n \Delta C.$$

(13)

High resolution measurements of the FAC rate downstream of a ferrule or orifice suggested that the rate did not increase with time as the surface roughened, Figure 27, and roughness effects do not appear to have been incorporated into the CEGB model.

The EDF model [65, 69, 72–76] has been developed with a number of modifications, most importantly the incorporation of the oxide porosity θ, oxide thickness δ, and k the rate constant for magnetite dissolution as important parameters. The original formulation in 1980 was

$$\text{FAC}_{\text{rate}} = K\,(C - Co)\,,$$

$$\text{FAC}_{\text{rate}} = 2k\,(Ceq - C)\,.$$

$$(14)$$

From which one gets

$$\text{FAC}_{\text{rate}} = \left\{\frac{2kK}{2k + K}\right\}(Ceq - Co)\,.$$

$$(15)$$

Then this was modified, after the Sanchez Calder model to take account of oxide porosity:

$$\text{FAC}_{\text{rate}} = \frac{\theta \Delta C}{[1/k + 0.5\,(1/K + \delta/D)]}\,.$$

$$(16)$$

And if mass transfer is the rate controlling step $[1/k \ll 0.5(1/K+\delta/D)]$ and for thin oxides $(1/K \gg \delta/D)$ this reduces to

$$\text{FAC}_{\text{rate}} = 2K\theta\Delta C.$$

$$(17)$$

The solubility of magnetite is obtained from Sweeton and Baes. Mass transfer coefficient is obtained by using a straight tube correlation modified for surface roughness and an enhancement factor to account for the component geometry:

$$K_{\text{component}} = 0.0193\left[\frac{\varepsilon}{d}\right]^{0.2} Re\ Sc^{0.4}\left(\frac{D}{d}\right)$$

$$\times \text{ component enhancement factor.}$$

$$(18)$$

Early versions of their model appeared to explain the variation of FAC with temperature and alloying effects, particularly the effect of chromium, by varying the values of oxide porosity θ and thickness δ. In the latest description of the model, the following is stated [65].

The porosity behavior with temperature is difficult to model. It appears to be temperature dependent and not linear. Based on many experiments using the CIROCO test loop, taking into account different temperatures, mass transfers and chemicals, (16) has been simplified as

$$FAC_{rate} = 2K \cdot f(Cr) \cdot H(T) \cdot \Delta C,$$

(19)

Where H (T) is a temperature-dependent function normalized to K and which leads to a hydrogen content (in $mg\,kg^{-1}$) equivalent for increasing the iron solubility. Only experiments can lead to this function and data are not shown.

The Sanchez Calder model [77] was developed to predict the rate of FAC in steam extraction lines. It incorporated most of the processes shown in Figure 26, with an equation similar to the EDF model, and highlighted their view of the importance of the oxide porosity, which was subsequently incorporated into some other models.

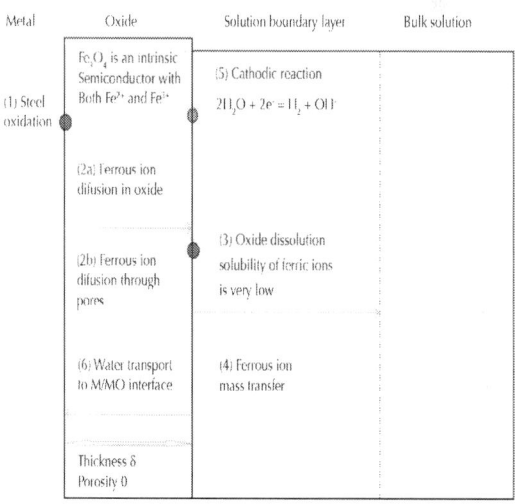

Figure 26: Schematic diagram of possible reactions in FAC.

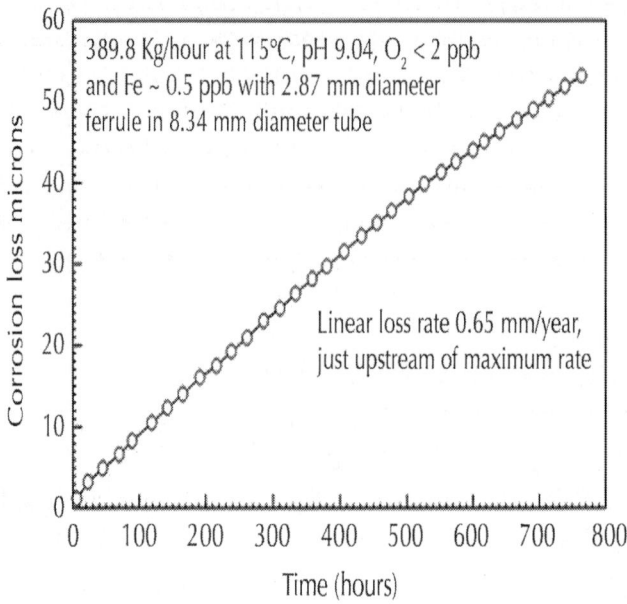

Figure 27: Linear FAC loss rate at orifice specimen ([78]).

Workers at Penn state Uni [79] pointed out the basic Sanchez Calder model predicted rates that were 100 to 1000 times smaller than measured rates and suggested changes to the model including:

- improved magnetite solubility data,
- assumption that porosity was a function of oxygen content of water,
- introduction of several piping configurations and surface finishes.

Since the occurrence of FAC at the outlet headers in the carbon steel primary circuit of CANDU reactors and the failure of a condensate pipe downstream of an orifice at Mihama PWR, various other mechanistic approaches have been formulated.

Burrill and Cheluget [80] followed a similar approach to CEGB workers in having the magnetite solubility (calculated using S&B) as a function of potential and thus the corrosion rate. However under CANDU condition of high pH and high temperature the dependency of magnetite solubility could be less than the normally accepted $[H]^{1/3}$ relationship.

Canadian/Japanese Approach [4, 15, 32, 81, 82]. Lister and Long suggested that an earlier model [83] which incorporated both film dissolution and mass transfer could not explain the occurrence of FAC on CANDU outlet feeders. This was because the film dissolution step was not sufficiently fast to keep up with the observed FAC rates. In addition because the FAC rate was a function of the velocity to a power of 1.5–2 it was stated that the surface shear stress τ must be important in causing magnetite removal. The stages in this model are thus as follows.

- Some magnetite dissolution occurs.
- Dissolution loosens magnetite crystals.
- Surface shear stress spalls or erodes crystals.
- Thinned magnetite is less protective.

An electrochemical model was developed which could predict corrosion potentials magnetite solubility's and FAC rates.

Subsequent work in collaboration with Japanese workers, including Uchida, has subsequently developed their own evaluation program because Unfortunately, details of their (other models) theoretical basis and of the data bases of the program packages are classified due to intellectual property rights. Both traceability of the computer program package and its validation are required for making policy on plant reliability when applying the package calculations.

A paper covering the evaluation of FAC simulation code based on verification and validation has been published [82], which describes the steps involved as follows.

Step 1: Define flow velocities and temperatures in system: 1D CFD code RELAPS used.

Step 2: Calculate O_2 levels ECP, and so forth: 1D O_2-hydrazine reaction code CORRENV used.

Step 3: Distribution of mass transfer coefficients: 3D CFD code PLASY MTCEXTRA used.

Step 4: Danger zone evaluation: chart analysis DRAWTHREE-FAC.

Step 5: Distribution of wall thinning: wall thinning code WATH.

Step 6: Evaluation of residual life and effect of counter measures: final evaluation DRAWTHREE-FAC.

Interestingly this paper appears to indicate the importance of obtaining the mass transfer rate but does not mention the importance of the surface shear stress.

Empirical Models

The Keller model [84–86] was probably one of the first predictive models. The basis for the formulation was a combination of experience gained from damage in wet steam systems and the pressure drops expected at various components:

$$\text{FAC}_{\text{rate}} = \left[f\left(T\right) f\left(X\right) V K_c \right] - Ks,$$

(20)

Where K_c is a component geometry factor and Ks is a constant thatmust be exceeded for FACto occur, is the steamquality, T is the temperature, and V is the velocity.

This was clearly inadequate since no environmental influence was included and was subsequently modified by Kastner et al. [85] to include the following.

- Environmental factors.
- A modified $K_{c'}$, which also included downstream influences:

$$\text{FAC}_{\text{rate}} = K_c \, f_1 \left(VT \text{ material composition}\right)$$

$$\times f_2\left(\text{pH}\right) f_3\left(O_2\right) f_4\left(\text{time}\right) f_5\left(X\right).$$

(21)

After the Surry failure this was developed into WATHEC and is used in conjunction with DASY to manage data obtained from NDT examination.

EPRI CHEC [2, 87, 88] (Chexal and Horriwtz). After the Surry failure EPRI collected all British, French, and German experimental data together with USA plant data. Other than being consistent with mechanistic understanding no presumptions were made as to the form of the correlation between FAC rate and all the influencing variables resulting in the formulation:

$$\text{FAC}_{\text{rate}} = f_1\left(T\right) f_2\left(\text{material composition}\right) f_3\left(K\right) f_4\left(O_2\right)$$

$$\times f_5\left(\text{pH}\right) f_6\left(\text{component geometry}\right) f_7\left(\alpha\right),$$

(22)

Where α is a factor for void fraction in two-phase flow. There have been various modifications and improvements made to this code and its use in plant examination.

Other Models

A trained artificial neural network (ANN) was used [79] to make predictions, but the details of the results were only given in graphical form. The following was suggested.

The combination of a deterministic model which reflects a mechanistic picture of FAC and a purely nondeterministic ANN model which reflects best experimental representation of the phenomenon are the best tools for extracting information from complex phenomenon such as FAC [79].

A Russian program ECW-02 based on CHECWORKS has been developed, and a Ukrainian program KASESC based on WATHEC has also been produced. RAMEK is a more original Russian approach that has been described and reviewed [8] that appears to combine a loss rate due to mass transfer with a loss rate due to the surface shear stress peeling of the oxide layer.

Various models [89–91] have been developed to predict droplet attack at bends in two-phase flow, but they appear to ascribe damage to be mechanical in nature. As outlined earlier the conditions causing such attack, as opposed to corrosion, need defining.

The relative usage of the various programs is that CHECWORKS is used in USA, Canada, Taiwan, Japan, South Korea, Czechia, Slovakia, and Slovenia ~153 units, EDF is only used in France but in all 58 units, and COMSY is used in Germany, Spain, Finland, Hungary, and Bulgaria ~17 units.

DISCUSSION

A simplified summary of how the prediction schemes deal with some of the variables that have been dealt with in this review is given in Table 5, and suggested reasons for the bell shaped dependency of FAC rate with temperature are summarized in Figure 29.

Table 5: Models and how they deal with key variables

Model	FAC function of V^m or K^n	Roughness effects	Solubility relationship	Cause of FAC reduction at Low T and High T
CEGB [54, 66, 67, 70, 71]	Nonlinear n>1<3	Acknowledged but not specifically integrated.	S and B	MDR* Solubility
EdeF [65, 69, 72–76]	Linear or less	Yes and specified relationship	S and B	Porosity Porosity MDR
Sanchez Calder [77]	Linear n=1or less	Apparently not	S and B	Porosity Porosity MDR
Penn S [79]	Mechanistic model-linear? ANN model nonlinear.	Yes but not specified	Their own which is bell shaped at pH 7 and increases with temperature at pH 9	Unclear
EPRI [2, 87, 88]	Not stated but m apparently between 0.6 and 0.7	Yes but suggested and did not occur	Unclear which or if incorporated into model, though clearly identified differences between S and B and T and LeB data	MDR Solubility
Siemens KWU [84–86]	FAC_{rate} e^{cv} where c is complex f of time and temperature	Apparently not	Unclear which or if incorporated into model	Unclear
Lister and Lang [15] and Uchida et al [81]	Nonlinear n>1due to shear stress on oxide	Yes	S and B or T and LeB?	Unclear

*MDR is magnetite dissolution reaction.

Importance of Mass Transfer

It is widely agreed that the mass transfer coefficient is the hydrodynamic parameter that controls the occurrence and rate of FAC. A relationship between rate of FAC and the mass transfer coefficient, with a power dependency of greater than one, is an erroneous reason to invoke the importance of τ the surface shear stress. It is believed that there are more cogent reasons to reject the involvement τ as an important parameter. These have been dealt with in detail earlier and elsewhere. Briefly the variations in τ, K, turbulence level, and flow assisted corrosion rate with distance downstream of an orifice are sufficient to confirm that τ is not important. It is probable that a similar statement will be found for other regions where detached flow occurs and the relationship between mass transfer heat-transfer and pressure drop breaks down.

Relationship between FAC and Mass Transfer and Magnetite Solubility

It is clear that there is not always a simple linear relationship between the FAC rate and the mass transfer rate, for example, the variation in FAC downstream of an orifice. Of the various predictive schemes only the EdF model appears to suggest a linear model, and the position of the EPRI code is unclear. Figure 28compares some CEGB and EDF data.

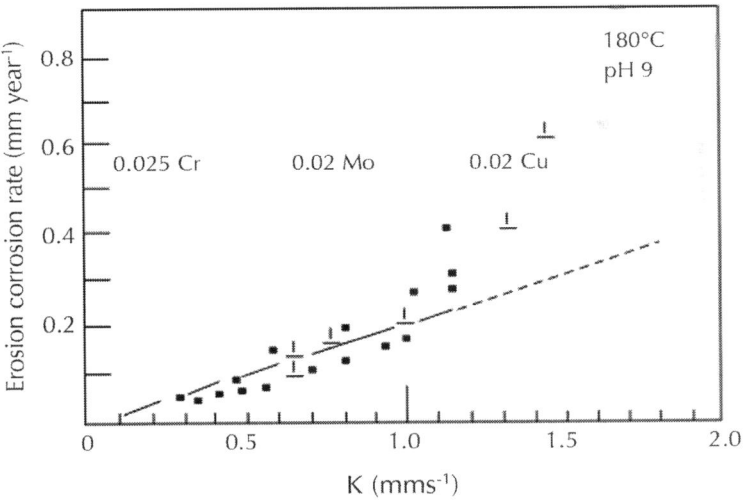

• 8 mm diameter tube
⊥ 5 mm impinging jet

(a)

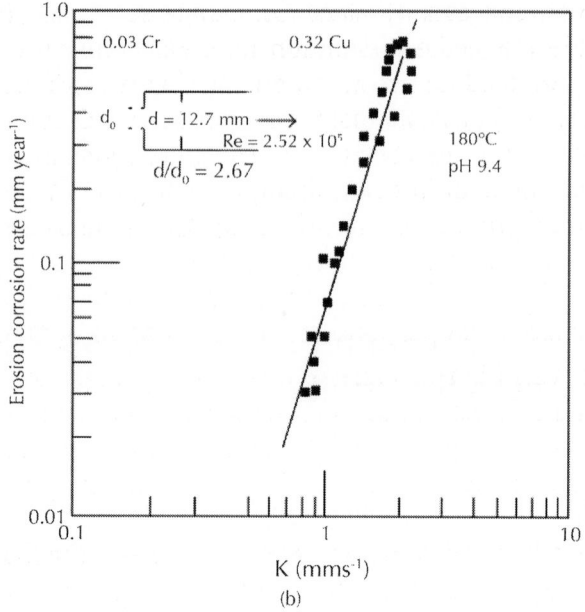

Figure 28: Suggested correlations between K and FAC rate. (a) French data ([74]). (b) British data ([12]).

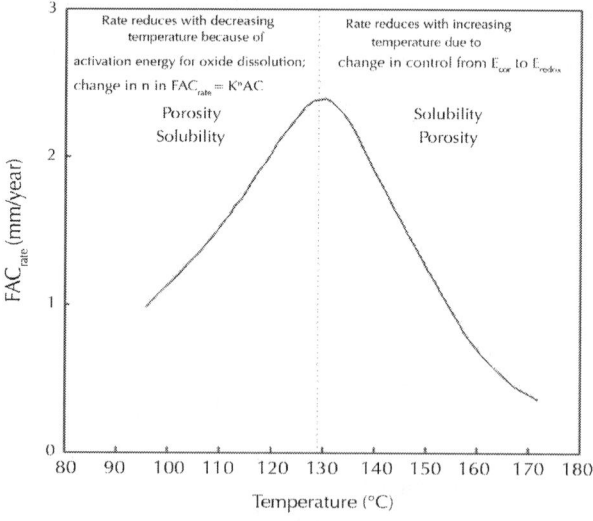

Figure 29: Effect of temperature on FAC with possible reasons.

It has been suggested that at low temperatures the FAC rate is limited by the magnetite dissolution rate. However an empirical fit of the data produced an activation energy that was a function of the mass transfer rate [70]. At temperatures above the FAC peak it has been suggested that redox reactions could become more important than the corrosion potential in determining the rate of FAC. Both of these situations would produce a lower dependency on mass transfer than at the peak temperature where the highest value of n is associated with the maximum feedback between the FAC rate, the corrosion potential, the production of hydrogen, and the solubility of magnetite.

Except for very high plateau FAC rates at low pH's (Figure 30) there is no convincing evidence for the importance of any processes other than mass transfer and magnetite solubility. The importance of magnetite dissolution kinetics being limiting at low temperatures appears to result from the widespread use of the solubility data of Sweaton and Baes. Most importantly this data suggests that magnetite solubility increases with decreasing temperatures down to about 60°C. However the data of Tremaine and LeBlanc show a bell shaped magnetite solubility curve (Figure 19), which more closely resembles the temperature dependence of FAC. In addition there are theoretical reasons why solubility should not be related to hydrogen partial pressure at low temperatures. Other reasons proposed to explain the temperature dependence of the FAC rate are summarized in Figure 28.

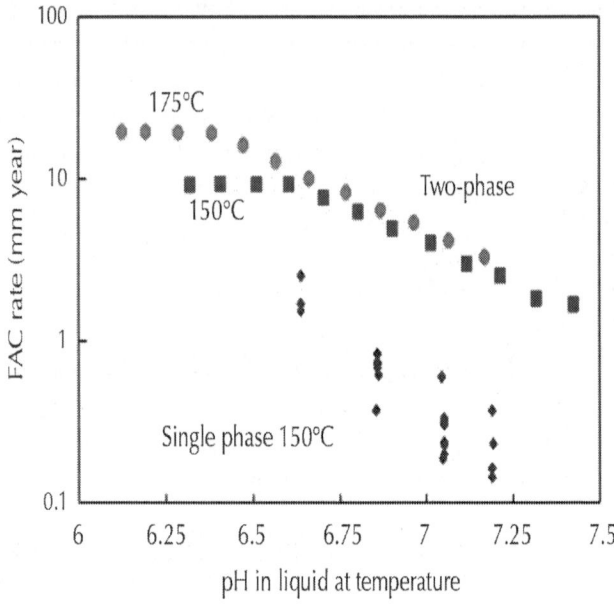

Figure 30: Effect of pH on FAC rate ([60]).

Influence of Roughness Development

The way the different models deal with roughness development, mass transfer, and the influences on the rate of FAC is summarised in Table 5. It appears that only the WATHEC model does not incorporate roughness effects. However it is not clear how the smooth to rough surface transition is handled by any of the models. The EdF model incorporates a formulation very similar to Figure 12 with the factor 0.01

being replaced with $0.193(\varepsilon / d)^{0.2}$. This would appear a reasonable formulation in line with pressure drop formulations. However it is clear that that as Re increases the size of scallops produced by FAC decreases, although the mass transfer enhancement over a smooth surface is increased; this appears inconsistent, except possibly during the initial stages of attack.

Mechanistic Comments on Prediction Models

It is surprising that after over 40 years of investigation and development there is no widespread agreement on a number of important mechanistic aspects, which have an impact on the ability to predict the rate of FAC, these include the following.

- The relevance of various processes particularly oxide porosity.
- The relationship between FAC and K
- Which solubility data to use
- The mechanism of improvement due to Cr content of the steel
- The importance and treatment of roughness development. Thus despite the ability to prevent its occurrence to say that FAC is well understood is somewhat premature.

Relationship of FAC to Deposition and SCC?

The roughness that develops during both FAC and the deposition of rippled magnetite both increase the resistance to flow. Such effects are usually represented graphically in terms of the friction factor (f) as a function of the Reynolds number; the friction factor relates the pressure drop and thus τ to the velocity. Such a diagram was constructed, by Moody for pipes having roughness elements of varying heights and distributions. Similar diagrams can be constructed for surfaces with uniform roughness, for example, with sand grains. There are some differences between these two types of diagrams [18]. However in both cases under turbulent conditions the friction factor increases with

the relative roughness (ε/d) and decreases with increases in Reynolds number. A simplified equation was recommended [18] to estimate such effects:

$$\frac{1}{f^{0.5}} = -3.6 \log_{10} \left\{ \frac{6.9}{Re} + \left(\frac{\varepsilon}{3.71d} \right)^{1.11} \right\}.$$

(23)

For surfaces that roughen as a result of dissolution or deposition the relative roughness produced decreases with increasing flow or Re. The evidence is that such roughness produced a greater resistance than expected from its (ε/d) value. From the upper bound mass transfer correlation and the analogy with heat and momentum transfer a single upper bound value of f can be estimated as a function of Re as f=0.02Re, but this has not been tested. It has been suggested, and there is some evidence for, the resistance peaking at the Reynolds number at which the roughness was formed [50, 92].

The difficulties predicting the occurrence of deposition have been indicated [64], although a velocity below 1.6–2.9 m/s was stated [93] as necessary to prevent the occurrence in straight tubes; otherwise pressure drop increases of 20 times at a Re of 10^6 could occur. Deposition appears preferentially at regions of high mass transfer and it has been suggested [94] that this is due to electrokinetically generated currents at regions of separated flow, but deposition does occur on straight pipes and bends, Figure 31. If electrokinetically currents are important deposition might be expected to occur at any dissolved iron levels (like electrodeposition). However there is still some debate about the roles of soluble or particulate iron. Like FAC deposition can be controlled by the addition of oxygen to the water or pH changes reducing the iron solubility.

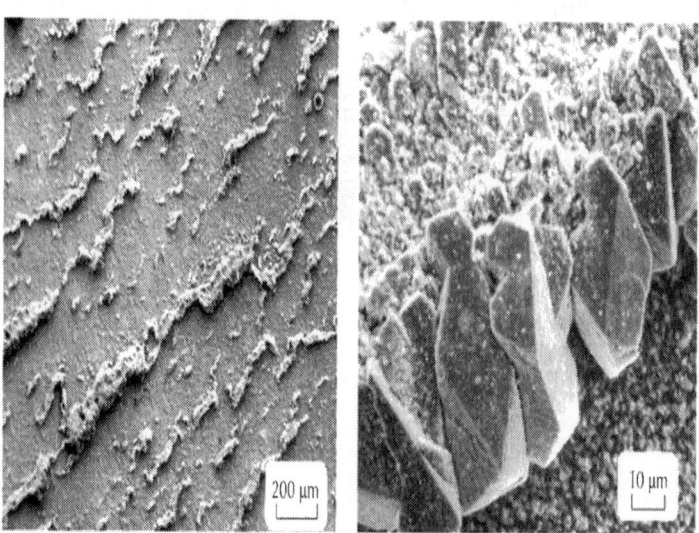

Figure 31: Typical magnetite deposit.

The SCC of highly stressed experimental orifice holders undergoing FAC was reported [12]. It was suggested that the electrochemical conditions for FAC and SCC are often similar. For example it was shown that the potential range promoting SCC and FAC, of carbon steels in carbonate solution, is identical. Also both SCC and FAC of carbon steels occur in carbonate, nitrate, and caustic solutions. The relationship between oxide solubility and crack propagation has also been suggested [95]. However from a practical viewpoint for carbon steels in water it is clear that cracking is favored by oxidizing conditions, which inhibit the occurrence of FAC. The occurrence of both FAC and SCC assumed importance by the occurrence of both FAC and SCC on the outlet headers of a CANDU plant; as yet there does not seem to be a full understanding of such cracking.

PREVENTING THE OCCURRENCE OF FAC

If the three major influences on FAC are considered, that is, material, environmental, and hydrodrodynamic, as shown in Figure 2, then how to prevent its occurrence is straightforward and depends if at the design stage or post construction and into operation.

The history of the British AGRs once through steam generators illustrates this rather well. Damage was first observed at and downstream of the flow control orifice assemblies situated in the feedwater inlet headers. These are accessible and the problem was corrected by a redesign which included a stainless steel section of tubing downstream of the orifice. However this problem highlighted the risk to the carbon steel bends in the economizer section of the boiler, which was controlled initially by raising the pH. Although this appeared successful it limited the life of ion exchange beds. It was known that oxygen additions to the feedwater could be beneficial in terms of general corrosion [96, 97]. A large program demonstrated that adding a small amount of oxygen prevented FAC [78, 98]. High resolution surface activation measurements of FAC rate confirm this as shown in Figure 32. In addition it was shown that additions of oxygen and hydrazine could be made together and that the oxygen would inhibit FAC in the economizer prior to being reduced by the hydrazine and thus minimise the possibility of SCC at oxidising potentials. Thus very cleverly two additions can be made to the secondary water, having opposite effects and preventing two forms of corrosion occurring in different regions of the boiler.

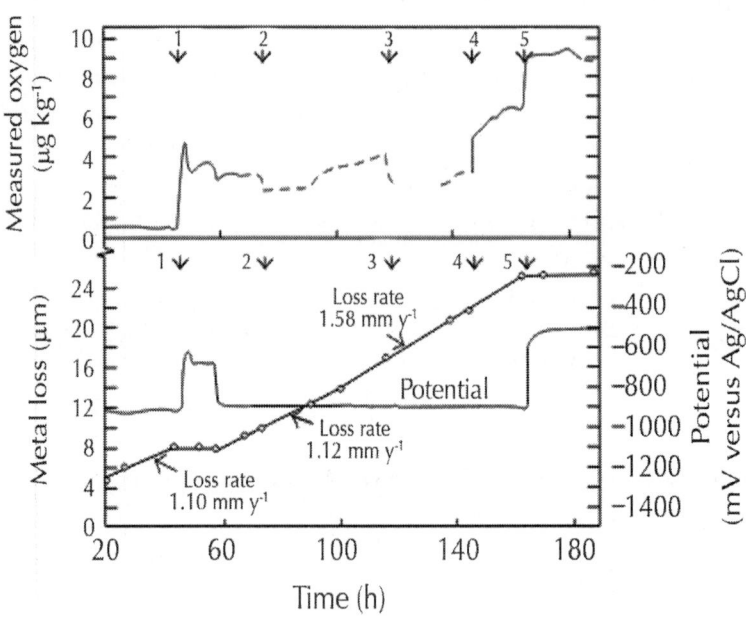

Figure 32: Effects of oxygen level on FAC rate and potential ([78]).

It was postulated that the amount of oxygen that was needed to be added to prevent FAC was related to the FAC rate by Faradic equivalence:

$$\text{Concentration of } O_2 > \frac{FAC_{rate}}{KO_2 \times P},$$

(24)

where P is a constant as defined in Figure 33. Note that the concentration (weight/volume) of oxygen is given by $O_2 \text{ppm} \times \rho_{water}$ the slope of the line in Figure 33 is a function of five parameters and its value seems to be an unreliable way of obtaining mechanistic information about the reactions involved.

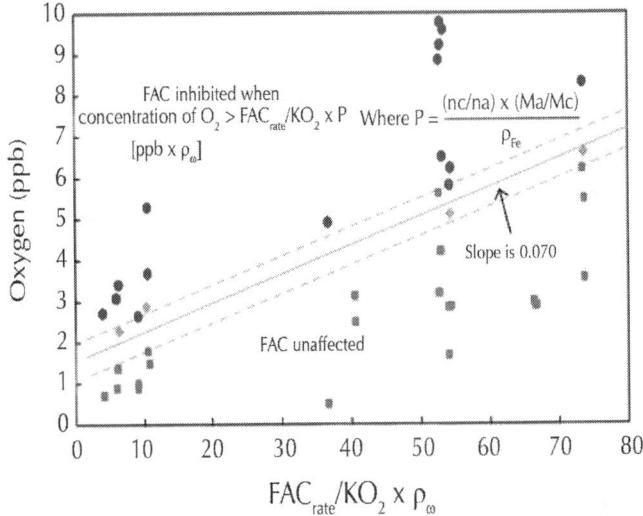

Figure 33: Suggested correlation for oxygen effect at 150°C (after [78]). nc, na and Ma, Mc are the number of electrons and the molecular weights of the anodic and cathodic species involved in the charge transfer steps, ρ_{Fe} is the density of iron, and ρ_w is the density of water.

The success of such oxygen additions has been described [99] and the approach has been adopted in other situations [100]. Later AGR boilers were constructed using a 1%Cr 0.5%Mo steel after extensive testing had confirmed its suitability. The use of 1%Cr 0.5%Mo steel has the advantage that no postweld heat-treatment is usually required unlike 2.25%Cr 1%Mo and higher alloyed steels.

CONCLUSIONS

- The available evidence continues to support the use of the mass transfer coefficient (K) as the best hydrodynamic parameter to characterize FAC in power plant.
- The fundamental relationship between K and the rate of flow accelerated corrosion continues to be a significant difference between the different predictive schemes. However, there is strong evidence that the rate of FAC is not always a simple linear function of the mass transfer rate.

- With normal flow as roughness develops mass transfer increases and the rate of FAC increases. There is now more evidence that this is much less important with detached flow.

- It is clear that the influence of the environmental parameters such as temperature, pH, and oxygen content is through their effect on the solubility of magnetite. However, the details of this and the variation of magnetite solubility deserve closer attention. In particular it might prove possible to explain the temperature dependency of FAC in terms of magnetite solubility and its relationship to hydrogen partial pressure, without having to invoke a change in the rate controlling step.

- There is the possibility that at higher pH's and temperatures, for instance those in the CANDU primary system the effect of hydrogen partial pressure, on magnetite solubility might not be proportional to $[H_2]^{1/3}$. As pH increases, at high temperatures, there will be a gradual reversal in the effect of hydrogen content on solubility, from being proportional to $[H_2]^{1/3}$ to. $[H_2]^{-1/6}$

- The beneficial effect of chromium continues to be demonstrated under all conditions. However, it is not clear if this effect is due to a reduction in the solubility of the magnetite, a change in the kinetics of oxide dissolution or its influence on the oxide porosity.

- It is clear that oxygen additions can prevent FAC but such additions to the environment might lead to other problems.

- There is still no agreement on a number of important mechanistic aspects. The mechanistic models appear to have adjustable parameters particularly the f(T) in the EDF code and the n in the K^n dependency of the FAC rate, in the CEGB model.

- The pragmatic models do not give enough information to completely understand their operation.

REFERENCES

1.   in Proceedings of the Conference on Corrosion-Erosion of Steels in High Temperature Water and Wet Steam, P. H. Berge and F. Khan, Eds., Electricite de France, Paris, France, 1983.

2.   "Flow-accelerated corrosion in power plants," TR-106611, EPRI, 1996.

3.   G. J. Bignold, C. H. de Whalley, K. Garbett et al., "Erosion-corrosion of mild steel ammoniated water," in Proceedings of the 8th International Congress on Metallic Corrosion, pp. 1548–1554, Mainz, Germany, 1981.

4.   S. Uchida, "Evaluation method for FAC of components by corrosion analysis coupled with flow dynamics analysis," in Proceedings of the Annual Meeting of the Executive Committee and Working Groups of the International Association for the Properties of Water and Steam (IAPWS '06), Witney, UK, September 2006.

5.   B. Poulson, B. S Greenwell, B. Chexal, and G. Horowitz, "Modelling hydrodynamic parameters to predict flow assisted corrosion—water reactors," in Proceedings of the 5th International Conference on Environmental Degradation of Materials in Nuclear Power Systems, American Nuclear Society, La Grange Park, Ill, USA, 1992.

6.   J. P. Slade and S. T. Gendron, "FAC and cracking of carbon steel piping in primary water-operating experience at the Point Lepreau Generating station," in Proceedings of the 12th International Conference on Envirinmental Degradation of Materials in Nuclear Power Systems-water Reactors, T. R. King, P. J. King, and L. Nelson, Eds., pp. 773–784, 2005.

7.   G. Cragnolina, "A review of erosion-corrosion of steels in high temperature water," in Proceedings of the 3rd Environmental Degradation of Materials in Nuclear Power Systems—Water Reactors, G. J. Theus and J. R. Weeks, Eds., pp. 397–406, The Metallurgical Society, 1988.

8.   I. Betova, M. Bojinov, and Saario, "Predictive modeling of flow-accelerated corrosion-unresolved problems and issues," VTT research report No VTT-R-08125-10, 2010.

9.   Y. S. Garud, "Issues and advances in the assessment of flow accelerated corrosion," Proceedings of the 14th International Conference on Environmental Degradation, 2008.

10.   G. J. Bignold, "Erosion-corrosion: history causes and remedies," Modern Power Systems, vol. 25, no. 2, pp. 11–15, 2005.

11.   B. Poulson, "Electrochemical measurements in flowing solutions," Corrosion Science, vol. 23, no. 4, pp. 391–430, 1983. · ·

12.   B. Poulson, "Predicting the occurrence of erosion corrosion," in Plant Corrosion: Prediction of Materials Performance, J. E. Strutt and J. R. Nichols, Eds., pp. 101–132, Ellis Harwood, Chichester, UK, 1987.

13.   B. Poulson, "Advances in understanding hydrodynamic effects on corrosion," Corrosion Science, vol. 35, no. 1–4, pp. 655–661, 1993. · ·

14.   B. Poulson, "Erosion-Corrosion," in Corrosion, L. L. Shrier, R. A. Jarman, and G. T. Burnstein, Eds., chapter 1–11, Butterworth/Heinnmann, Oxford, UK, 3rd edition, 1994.

15.   D. H. Lister and L. C. Lang, "A mechanistic model for predicting FA and GC of carbon steel in reactor primary coolants," in Proceedings of the International Conference on Water Chemistry of Nuclear Reactor Systems (CHIMIE ‹02), Societe Francaise d›Energie Nucleaire, Avignon, France, 2002.

16.   D. C. Silverman, "RCE-an approach for predicting velocity sensitive corrosion," in Flow Induced Corrosion: Fundamental Studies and Industrial Experience, K. J. Kennelley, R. H. Hausler, and D. C. Silverman, Eds., NACE, Houston, Tex, USA, 1992.

17.   G. F. Hewitt, Measurement of two phase flow parameters, Academic Press, London, UK, 1978.

18.   B. Massey, Mechanics of Fluids, Taylor & Francis, London, UK, 8th edition, 2006, Revised by J. Ward-Smith.

19.   A. A. Wragg, "Applications of the limiting diffusion current technique in chemical engineering,"Chemical Engineer, no. 316, pp. 39–49, 1977.

20.   T. Mizushina, "The electrochemical method in transport phenomena," Advances in Heat Transfer, vol. 7, pp. 87–161, 1971.

21.   B. Poulson and R. Robinson, "The use of a corrosion process to obtain mass transfer data," Corrosion Science, vol. 26, no. 4, pp. 265–280, 1986. · ·

22.   M. W. E. Coney, S. J. Wilkin, and H. S. Oates, Thermal-hydraulic effects on mass transfer behaviour and on erosion corrosion metal loss rate paper 13 in ref 1, 1983.

23.   D. D. Wang, "Characterization of local mass transfer rate downstream of an orifice," Open Access Dissertations and

Theses. Paper 7142, 2012, http://digitalcommons.mcmaster.ca/opendissertations/ 7142.

24. A. Etebari, "Recent innovations in wall shear stress sensor technologies," Recent Patents on Mechanical Engineering, vol. 1, no. 1, pp. 22–28, 2008.

25. S. M. Chouikhi, M. A. Patrick, and A. A. Wragg, "Two phase turbulent wall transfer processes downstream of abrupt enlargements of pipe diameter," in Proceedings of the International Conference on the Physical Modelling of Multi-Phase Flow Coventry, BHRA, Cranfield, UK, 1983.

26. G. Schmitt, C. Werner, and M. Bakalli, "Fluid mechanical interactions of turbulent flowing liquids with the wall-revisited with a new electrochemical tool," Paper 5344, NACE, Houston, Tex, USA, 2005.

27. G. Schmitt and M. Bakalli, "Maximum flow intensities at tools for measuring flow influenced corrosion," Paper 9472, NACE, Houston, Tex, USA, 2009.

28. G. Schmitt and M. Bakalli, "A critical review of measuring techniques for corrosion rates under flow conditions," PowerPlant Chemistry, vol. 9, no. 6, pp. 89–106, 2007.

29. S. Nesic and J. Postlethwaite, "Hydrodynamics of disturbed flow and erosion-corrosion part I: single-phase flow study," The Canadian Journal of Chemical Engineering, vol. 69, no. 3, pp. 698–702, 1991. · ·

30. G. Schmitt and T. Gudde, "Local mass transport coefficients and local wall shear stresses at flow disturbances," paper 102 at Corrosion, NACE, Houston, Tex, USA, 1995.

31. I. S. Woolsey, Private communication, 1983.

32. D. H. Lister and S. Uchicia, "Reflections on FAC mechanisms," Power Plant Chemistry, vol. 12, no. 10, pp. 590–597, 2010.

33. M. Matsumura and Y. Oka, "Mechanism of erosion corrosion of copper alloys," in Proceedings of the Pacific Corrosion, Melbourne, Australia, 1987.

34. M. Matsumura, Y. Oka, S. Okumoto, and H. Furuya, "Jet-in-slit test for studying erosion corrosion,"STP866, ASTM, 1985. ·

35. M. Coney, CERL Report RD/L/N197/80, 1980.

36. G. Schmitt and M. Mueller, Critical Wall Shear Stresses in CO_2 Corrosion of Carbon Steel, Paper 44 Corrosion 99, NACE, Houston, Tex, USA, 1999.

37. B. Poulson and R. Robinson, "The local enhancement of mass transfer at 180° bends," International Journal of Heat and Mass Transfer, vol. 31, no. 6, pp. 1289–1297, 1988. · ·

38. B. Poulson, "Mass transfer from rough surfaces," Corrosion Science, vol. 30, no. 6-7, pp. 743–746, 1990. · ·

39. B. Poulson, "Measuring and modelling mass transfer at bends in annular two phase flow," Chemical Engineering Science, vol. 46, no. 4, pp. 1069–1082, 1991. · ·

40. B. Poulson, B. S. Greenwell, B. Chexal, and G. Horowitz, "Roughness effects on flow assisted corrosion," in Proceedings of the International Conference on Interaction of Iron Based Materials with Water and Steam, B. Dooly and A. Bursik, Eds., EPRI Palo Alto EPRI TR 102102, 1993.

41. B. Poulson, "Complexities in predicting erosion corrosion," Wear, vol. 233–235, pp. 497–504, 1999. · ·

42. S. Uchida, M. Naitoh, Y. Uehara, H. Okada, and D. H. Lister, "Evaluation method for FAC of components by corrosion analysis coupled with flow dynamics analysis," in Proceedings of the 13th International Conference on Environmental Degradation of Materials in Nuclear Power Systems, CNS, Whistler, Canada, August 2007.

43. E. Hoashi, S. Yoshihashi-Suzuki, T. Kanemura, et al., "Development of 3D CFD technology for flow-assisted corrosion," in Proceedings of the Conference on Water Chemistry of Nuclear. Reactor Systems (NPC ‹08), Berlin, Germany, 2008.

44. J. M. Pietralik and B. A. W. Smith, "CFD application to FAC in feeder bends," in Proceedings of the 14th International Conference on Nuclear Energy (ICONE ‹06), Miami, Fla, USA, 2006.

45. J. M. Pietralik and C. S. Schefski, "Flow and mass transfer in bends under FAC wall thinning conditions," in Proceedings of the 17th International Conference on Nuclear Energy (ICONE ‹09), Brussels, Belgium, 2009.

46. D. Zinemams and A. Herszaz, "Flow accelerated corrosion: flow field and mass transport in bifurcations and nozzles," in

Proceedings of the International Conference on Water Chemistry for Nuclear Reactor Systems, NPC, Berlin, Germany, 2008.

47. K. Yoneda, "Evaluation of hydraulic factors affecting flow accelerated corrosion and its verification with power plant data," in Proceedings of the ASME 2009 Pressure Vessels and Piping Conference, PVP, Prague, Czech Republic, 2009.

48. J. R. L. Allen, "Bed forms due to mass transfer in turbulent flows: a kaleidoscope of phenomena,"Journal of Fluid Mechanics, vol. 49, pp. 49–63, 1971.

49. P. N. Blumberg and R. L. Curl, "Experimental and theoretical studies of dissolution roughness,"Journal of Fluid Mechanics, vol. 65, no. 4, pp. 735–751, 1974. · ·

50. R. M. Thomas, "Size of scallops and ripples formed by flowing water," Nature, vol. 277, no. 5694, pp. 281–283, 1979. · ·

51. L. Tomlinson and C. B. Ashmore, "Erosion corrosion of carbon and low alloy steels by water at 300 degree C," The British Corrosion Journal, vol. 22, no. 1, pp. 45–52, 1987. · ·

52. H. M. Crockett and J. S. Horowitz, "Low temperature FAC," in Proceedings of the ASME PV&P Div (PVP ‹09), ASME, Prague Czech Republic, 2009.

53. P. R. Tremaine and J. C. LeBlanc, "The solubility of magnetite and the hydrolysis and oxidation of ferrous ions in water to 300°C," Journal of Solution Chemistry, vol. 9, no. 6, pp. 415–441, 1980.

54. G. J. Bignold, "Erosion-corrosion of steels in feedwater—a broader application of the theory," inProceedings of the International Conference on Interaction of Iron Based Materials with Water and Steam, B. Dooly and A. Bursik, Eds., EPRI TR 102102, EPRI, Palo Alto, Calif, USA, 1993.

55. S. E. Ziemniak, M. E. Jones, and K. E. S. Combs, "Magnetite solubility and phase stability in alkaline media at elevated temperatures," Journal of Solution Chemistry, vol. 24, no. 9, pp. 837–877, 1995. · ·

56. F. H. Sweeton and C. F. Baes Jr., "The solubility of magnetite and hydrolysis of ferrous ion in aqueous solutions at elevated temperatures," The Journal of Chemical Thermodynamics, vol. 2, no. 4, pp. 479–500, 1970. · ·

57. G. Bohnsach, Solubility of Magnetite in Water and Aqueous Solutions of Acid and Alkalies, Vulkan, Essen, Germany, 1987.

58. S. M. Walker and E. W. Thorton, "Reanalysis of oxide solubility data," in Proceedings of the Water Chemistry of NRS, vol. 5, pp. 89–95, BNES, London, UK, 1989.

59. M. Bojinov, P. Kinnunen, K. Lundgren, and G. Wikmark, "Characterisation and modelling of oxide films on stainless steels and Ni alloys in light water reactors," VTT Report No. BTU073-041285, 2004.

60. G. J. Bignold, "Distribution of solutes between water and steam- influence on two phase," inProceedings of the FAC 8th International Conference Combined Cycle Plants with Heat Recovery Systems, EPRI Calgary, 2006.

61. M. Coney, "Erosion-Corrosion: the calculation of mass-transfer coefficients," CERL ReportRD/L/N197/80, 1980.

62. J. Newman, "Mass transport and potential distributions in geometries of localized corrosion," inProceedings of the U. R. Evans Conference on Localized Corrosion, R. W. Staehle, B. F. Brown, and J. Kruger, Eds., vol. 3, pp. 45–61, NACE, Houston, Tex, USA, 1974.

63. W. M. M. Huijbregts, "The influence of chemical compositionof carbon steel on erosion corrosion in wet steam," in Proceedings of the Conference on Corrosion-Erosion of Steels in High Temperature Water and Wet Steam, P. H. Berge and F. Khan, Eds., Electricite de France, Paris, France, 1983.

64. B. Poulson, H. Gartside, and R. Robinson, "Corrosion aspects associated with orifice plates and orifice assemblies›," in Proceedings of the Conference on Corrosion-Erosion of Steels in High Temperature Water and Wet Steam, P. H. Berge and F. Khan, Eds., paper no 10, Electricite de France, Paris, France, 1983.

65. S. Trevin, FAC in Nuclear Power Plant Components in Nuclear Corrosion Science and Engineering, Damien Féron Woodhead Publishing, Cambridge, UK, 2012.

66. G. J. Bignold, K. Garbett, R. Garnsey, and I. S. Woosey, "Erosion-corrosion in nuclear steam generators," INIS Collection Search, pp. 1–14, 1981.

67. J. G. Bignold, C. H. DeWhalley, K. Garbett, and I. Woolsey, "Mechanical aspects of erosion corrosion under boiler feedwater conditions," in Water Chemistry, vol. 3, pp. 219–226, British Nuclear Energy Society, London, UK, 1983.

68. K. A. Burrill, E. L. Cheluget, and M. Chocron, "Modelling the effect of Cr on FAC in reactor circuits," in Proceedings of the Water Chemistry of NRS BNES, pp. 528–537, Bournemouth, UK, 1999.

69. M. Bouchacourt, "Improvements in EC mechanistic model: role of surface film," in Water Chemistry of Nuclear Reactor Systems, vol. 6, pp. 338–340, BNES, London, UK, 1992.

70. G. J. Bignold, K. Garbett, and I. S. Woolsey, "Erosion corrosion in boiler feedwater; comparison of laboratory and plant data," in Proceedings of the UK Corrosion Conference, Institute of Corrosion Science and Technology, Birmingham, UK, 1983.

71. A. J. Bates, G. J. Bignold, K. Garbett et al., "CEGB single phase erosion-corrosion research programme," Nuclear Energy, vol. 25, no. 6, pp. 361–370, 1986.

72. B. J. Ducreux and P. Saint-Paul, "Effects of chemistry on erosion-corrosion of steels in water," in Proceedings of the Water Chemistry of NRS 2 BNES, pp. 19–23, London, UK.

73. J. Ducreux, "The influence of flow velocity on the corrosion-erosion of carbon steel in pressurized water," in Proceedings of Water Chemistry of Nuclear Reactor Systems, pp. 227–233, British Nuclear Energy Society, London, UK.

74. M. Bouchacourt and F.-N. Remy, EDF Study of FAC for the French PWR Secondary Circuit, Predicting the life of Corroding Structures, NACE, Cambridge, UK, 1991.

75. E.-M. Pavageau, O. de Bouvier, S. Trévin, J.-L. Bretelle, and L. Dejoux, "Dejoux Update of the water chemistry effect on the FAC rate of carbon steel: iinfluence of hydrazine, boric acid, ammonia, morphine and ethanolamine," in Proceedings 13th International Conference on Environmental Degradation of Materials in Nuclear Power Systems, CNS, Whistler, Canada, August 2007.

76. E. Ardillon, B. Villain, and M. Bouchacourt, "Probabilistic analysis of FAC in French PWR: the probabilistic version of BRT-CICERO version 2".

77. L. E. Sanchez-Caldera, P. Griffith, and E. Rabinowicz, "The mechanism of corrosion-erosion in steam extraction lines of power stations," Journal of Engineering for Gas Turbines and Power, vol. 110, no. 2, pp. 180–184, 1988. · ·

78. I. S. Woolsey, G. J. Bignold, C. H. de Whalley, and K. Garbett, "The influence of oxygen and hydrazine on E-C behaviour and electrochemical potentials of carbon steel under boiler feedwater conditions," in Proceedings of the 4th International Conference on Water Chemistry of Nuclear Reactor Systems, British Nuclear Energy Society, 1986.

79. M. Urquidi-Macdonald, D. V. Vooris, and D. D. Macdonald, "Prediction of single phase erossio corrosion in mild steel pipes using artificial neural networks and a deterministic model," Corrosion 95 paper 546 NACE, 1995.

80. K. A. Burrill and E. L. Cheluget, "Corrosion of CANDU outlet feeder pipes," in Proceedings of the JAIF International Conference on Water Chemistry, NPP, Kashiwazaki, Japan, 1998.

81. S. Uchida, I. M. Naito, Y. Uehara, H. Okada, S. Koshizuka, and D. H. Lister, Evaluation of Flow Accelerated Corrosion of PWR Secondary Components by Corrosion Analysis Coupled with Flow Dynamics Analysis, Springer, Berlin, Germany, 2008.

82. S. Uchida, M. Naitoh, H. Okada, T. Ohira, S. Koshizuka, and H. Derek, "Lister evaluation of FAC simulation code based on verification and validation," Power Plant Chemistry, vol. 12, no. 9, pp. 550–559, 2010.

83. W. G. Cook, D. H. Lister, and J. M. McInerney, "The effects of high liquid velocity and coolant chemistry on material transport in PWR coolants," in Water Chemistry of Nuclear Reactor Systems, vol. 8, BNES, London, UK, 2000.

84. H. Keller, "Erosionskorrosion in Nassdampfturbinen," VGB Kraftwerkstechnik, vol. 54, no. 5, pp. 292–295, 1974.

85. W. Kastner, M. Erve, N. Henzel, and B. Stellwag, "Calculation code for erosion corrosion induced wall thinning in piping systems," Nuclear Engineering and Design, vol. 119, no. 2-3, pp. 431–438, 1990. · ·

86. H. Nopper and A. Zanderepri, "Lifetime evaluation of plant components affected by FAC with the COMSY code," in Proceedings 13th International Conference on Environmental

Degradation of Materials in Nuclear Power Systems, CNS, Whistler, Canada, August 2007.

87. C. Schefski, J. Pietralik, T. Dyke, and M. Lewis, "Flow-accelerated corrosion in nuclear power plants: application of checworks at darlington," in Proceedings of the 3rd CNS International Conference on CANDU Maintenance, Toronto, Canada, November 1995.

88. H. M. Crockett and J. S. Horowitz, "Determining FAC degradation from NDE data," in Proceedings of the ASME Pressure Vessels and Piping Conference, pp. 909–918, Prague, Czech Republic, July 2009.

89. T. Ohira, R. Motira, K. Tanji, et al., "Prediction of liquid droplet impingement erosion (LDI) trend in actual NPP," in Proceedings of the AMSE PVP, Prague, Czech Republic, 2009.

90. R. Morita, F. Inada, and K. Yoneda, "Development of evaluation system for liquid droplet impingement erosion (LDI)," in Proceedings of the AMSE PVP 20009, Prague, Czech, 2009.

91. M. Satou, T. Sato, and A. Hasegawa, "Role of oxide layer on wall thinning caused by liquid droplet impingement (PVP ‹09)," in Proceedings of the AMSE Pressure Vessels and Piping Conference, Prague, Czech Republic, 2009. ·

92. W. Schoch, H. Wiehn, R. Richter, and H. Schuster, "Increase in pressure loss and magnetite formation in a benson boiler," British Corrosion Journal, vol. 6, pp. 258–268, 1971.

93. L. M. Wyatt, Materials of Construction for Steam Power Plant, Applied Science Publishers, London, UK, 1976.

94. M. I. Woolsey, R. M. Thomas, K. Garbett, and G. J. Bignold, "Occurrence and prevention of enhanced oxide deposition in boiler flow control orifices," in Proceedings of the 5th International Conference on the Water Chemistry of Nuclear Reactor Systems, vol. 1, pp. 219–228, BNES, London, UK.

95. J. R. Weeks, B. Vyas, and S. H. Isaacs, "Environmental factors influencing SCC in boiling water reactors," Corrosion Science, vol. 25, pp. 757–768, 1985.

96. R. K. Freier, "Protective film formation on steel by oxygen in neutral water free of dissolved solids," inProceedings of the VGB Feedwater Conference, pp. 11–17, 1969.

97. G. M. W. Mann, "The oxidation of Fe base alloys containing less than 12%Cr in high temperature aqueous solutions," in High Temperature High Pressure Electrochemistry in Aqueous Solutions: NACE-4, pp. 34–47, NACE, Houston, Tex, USA, 1976.

98. G. M. Gill, J. C. Greene, G. S. Harrison, D. Penfold, and M. A. Walker, "The effects of oxygen and iron in feedwater on erosion-corrosion of mild stel tubing paper 15," in Proceedings of the Conference on Corrosion-Erosion of Steels in High Temperature Water and Wet Steam, P. H. Berge and F. Khan, Eds., Electricite de France, Paris, France, 1983.

99. G. P. Quirk, I. S. Woolsey, and A. Rudge, "Use of oxygen dosing to prevent flow-accelerated corrosion in advanced gas-cooled reactors," Power Plant Chemistry, vol. 13, no. 4, 2011.

100. W. Ruhle, H. Neder, G. Holz, and V. Schneider, "Oxygen injection into reheater steam of moisture separator reheaters," PowerPlant Chemistry, vol. 7, no. 6, pp. 355–363, 2005.

Chapter 3

Investment Casting of Nozzle Guide Vanes from Nickel-based Superalloys: Part I – Thermal Calibration and Porosity Prediction

Agustín Jose Torroba[1], Ole Koeser[2], Loic Calba[2], Laura Maestro[3], Efrain Carreño-Morelli[1], Mehdi Rahimian[4], Srdjan Milenkovic[4], Ilchat Sabirov[4], and Javier LLorca[4, 5]

[1]University of Applied Sciences and Arts Western Switzerland, Sion, Switzerland

[2]CALCOM-ESI, Lausanne, Switzerland

[3]Precicast Bilbao, Bilbao, Spain

[4]IMDEA Materials Institute, C/Eric Kandel 2, Getafe 28906, Madrid, Spain

[5]Department of Materials Science, Polytechnic University of Madrid, Madrid, Spain

ABSTRACT

Investment casting is the only commercially used technique for fabrication of nozzle guide vanes (NGVs), which are one of the most important structural parts of gas turbines. Manufacturing of NGVs has always been a challenging task due to their complex shape. This work focuses on development of a simulation tool for investment casting of a new generation NGV from MAR-M247 Ni-based superalloy. A thermal model is developed to predict thermal history during investment casting. Experimental casting trials of the NGV are carried out and the thermal history of metal, mold, and insulation wrap is recorded. Inverse modeling of the casting trials is used to define accurately some thermophysical parameters and boundary conditions of the thermal model. Based on the validated thermal model, another model is developed to predict porosity in the as-cast NGVs. The porosity predictions are in good agreement with the experimental results in the as-cast NGVs. The advantages and shortcomings of the developed modeling tool are discussed.

BACKGROUND

Nozzle guide vanes (NGVs) are important structural parts of gas turbines [1]. NGVs are typically made from Ni-based superalloys because they have to withstand very high temperatures and aggressive environments [2]. Investment casting in vacuum, also often referred to as lost-wax process, is the only commercially used manufacturing route of these parts that have very complex shapes [3]. Large improvements in turbine efficiency can be achieved with improved designs of the NGVs that normally lead to more complex shapes and thinner geometries. However, these innovations are hindered by the complexity of the manufacturing process, which leads to an increasing number of defects (mainly porosity) during investment casting of parts with complex shapes and very thin elements. As a result, the development of investment casting routes for the new generation of NGVs is carried out via a 'trial and error' approach or, in other words, via experimental casting trials. But this strategy is very expensive and time consuming and thus dramatically limits the rate of innovation.

Presently, a paradigm shift is underway in which the experimental casting trials are partially replaced by the numerical simulation of the investment casting process to overcome the limitations of standard 'trial and error' approach [3]. Reconfiguration of the mold that was made on the foundry floor can now be made on a computer and simulated. In addition, the thermal history of a casting can be examined by means of simulations, and the effect of the casting parameters on the microstructure and quality of the as-cast parts can be evaluated. For example, Anglada et al. [4] and Rafique and Iqbal [5] successfully performed the simulation of heat transfer during investment casting of prototypes from Ni-based superalloys. A short description of the modeling tools developed to date and their application to casting of Ni-based superalloys is provided below.

Models Developed for Porosity Prediction

Porosity is known to be the most common defect found during investment casting and dramatically limits the strength and fatigue life of aerospace components [6]. Thus, investment casting foundries strive to minimize, if not eliminate, this insidious and persistent defect. The available modeling strategies for the prediction of porosity can be classified into three main groups, which are briefly described below.

Analytical Models

Computer models describing the formation of microporosity on the scale of the casting are based on volume-averaging methods for the calculation of the local temperature and pressure fields in the inter-dendritic liquid. These quantities are then used to estimate the level of gas segregation and to determine if the conditions for the nucleation of a pore are met. Most of these approaches originate from the pioneering work by Piwonka and Flemings [7], who developed analytical models that range from exact mathematical solutions to asymptotic approximations using 1D Darcy's law for pore nucleation. A constant solidification velocity together with a constant thermal gradient were assumed in these models. In order to obtain a more accurate prediction of the pore size, the gas pressure within the pores was included in the model, leading to a reasonable agreement with experimental results.

Criteria Function Models

Criteria functions were developed in the 1950's for dimensioning the size of risers and prevent inter-dendritic centerline shrinkage and porosity in steel plates [8]. Among the different criteria functions proposed, the Niyama criterion is the most widely used in metal casting to predict feeding-related shrinkage porosity caused by shallow temperature gradients [9]. The Niyama function N_y is a local thermal parameter defined as $N_y = G/\dot{T}$, where G is the local temperature gradient and $\dot{T}$ the local cooling rate. It is assumed that shrinkage porosity will form in regions in which the Niyama parameter is below a given threshold, which should be experimentally determined for each alloy. A dimensionless form of the Niyama function was presented in [10] that accounts for not only the thermal parameters but also the properties and the solidification characteristics of the alloy and it is able to predict the shrinkage pore volume fraction from the solid fraction-temperature curve and the total solidification shrinkage of the alloy.

Numerical Models

The first model for porosity prediction was developed by Kubo and Pehlke in two-dimensions (2D) [11] and was based on the relationship between the fraction of porosity and local pressure. Lee and Hunt [12] simulated the growth of pores due to hydrogen diffusion in Al-Cu alloys using a 2D continuum diffusion model, combined with a stochastic model of pore nucleation. Although the model did not include the effect of pressure drop due to shrinkage, it showed a good correlation with *in situ* observations of pore growth. Later, Lee et al. [13] developed a multi-scale model of solidification in Al-Si-Cu alloys, including microsegregation and microporosity. Pro CAST was used to solve the energy, momentum, and continuity equations to determine the temperature and pressure evolution with time. This information was coupled to a mesoscale model for microstructural development. Carlson et al. proposed a volume-averaged model for finite rate diffusion of hydrogen in Al alloys [14]. They coupled the calculation of the micro/macroscale gas species transport in the melt with a model for the feeding, flow, and pressure field. This was the first work considering hydrogen diffusion in the growth of pores for three-dimensional (3D) calculations. Pequet et al. developed a 3D microporosity model based on the solution of Darcy's equation and microsegregation of gas [15]. The model coupled microporosity with macroporosity and pipe-shrinkage predictions in a coherent way, with appropriate boundary conditions. Later, this approach was improved by developing a porosity model for multi-gas systems in multi-component alloys, including a realistic model for pore pinching [16], [17].

Porosity Prediction in Casting of Ni-based Superalloys

Most of the research on porosity prediction has been focused in Al alloys and steels [18],[19], and the work on Ni-based superalloys is more limited. Simulation of solidification to predict porosity in investment castings from Ni-based superalloys started a long time back; though, very simple geometries were considered in the earlier works. Overfelt et al. [20] developed a computer solidification model for the castings of plates with thicknesses of 2.54, 12.7, and 25.4 mm made from the In-718 Ni-based superalloy. The model was used to validate and disprove various phenomenological criteria for predicting porosity. The computer model was shown to be effective in predicting unfed centerline shrinkage in the 25.4-mm-thick plates, but it did not provide precise results for the thinner plates. Monastyrskiy [21] proposed a modeling tool based on liquid metal deformation due to solidification to model shrinkage porosity formation in a GS 32 Ni-based superalloy with low gas content. The model predicted the volume fraction and size of the shrinkage porosity. Nucleation of pores depended on the local stress level in the melt and the pore growth was driven by stress relaxation after pore nucleation. Numerical studies of directional solidification under an imposed temperature gradient and cooling rate were in good agreement with experimental data on porosity formation in Ni-based superalloys [21].

Modeling of investment casting of complex-shape parts have shown to be a more challenging task. Kang et al. [22] applied a model based on the dimensionless Niyama criterion to predict the formation of microporosity in a Ni-based superalloy casting containing complex shapes with thin walls. The theoretical predictions of microporosity showed reasonable agreement with the experimental results, though they underestimated the porosity content in the complex thin-wall regions. However, the model was not suitable for the shrinkage porosity prediction in the thick parts of the casting, since those sections often formed isolated liquid pools.

In this work, an advanced modeling approach is applied to the development of a new generation of NGVs with complex shape for aero engines. The objective of this work is twofold. Firstly, a thermal model capable of predicting the thermal history during investment casting of the new generation NGVs is developed. The principles of the thermal model were earlier described by Calba and Lefebvre[23]. Once the developed thermal model is validated against experimental results, the overall casting process can be analyzed in detail. The second aim of the present work is to simulate the development of defects in the as-cast NGVs (such as shrinkage porosity) and the final grain structure. The present manuscript consists of two parts and this (first) part focuses on the development and validation of the thermal model and the porosity prediction tool.

Material and Experimental Procedures

Investment casting of the NGVs was carried out using MAR-M247 Ni-based superalloy. The chemical composition of the material is presented in Table 1. The MAR-M247 superalloy is characterized by high temperature strength and excellent corrosion and oxidation resistance at elevated temperatures [24].

Table 1: The chemical composition of the MAR-M247 Ni-based superalloy

Ni	C	Cr	Co	Mo	W	Ta	Al	Ti	Hf
Base	0.15	8.4	10	0.7	10	3.1	5.5	1.05	1.4

Torroba *et al.*

Torroba *et al. Integrating Materials and Manufacturing Innovation* 2014 3:25   doi:10.1186/s40192-014-0025-5

The ceramic molds for the experimental casting trials were prepared using the standard manufacturing route. The wax pattern for the NGV was prepared via injection molding and then assembled with a wax feeding system. The obtained wax cluster was immersed into ceramic slurry and allowed to dry, and this step was repeated until the desired thickness of ceramic mold was reached. The assembly was dewaxed in an autoclave for 15 min at elevated temperature and high pressure. To burn the wax remains, the ceramic cluster was fired at 900°C for 1 h. Finally, the interior of the ceramic cluster was thoroughly rinsed.

The ceramic cluster was wrapped by an insulation layer (made from kaolin wool), having a thickness of 13 mm and was preheated to 1,200°C. The geometry and mesh for the model, including different cross sections, are presented in Figure 1b. Before entering the casting furnace, the thermocouples for recording the thermal history during investment casting were quickly set on the assembly. The equipment for *in situ* temperature measurements consisted of K- and S-type thermocouples and a standalone data logger able to withstand high vacuum (10^{-3} mbar), magnetic fields (coming from the induction furnace), and thermal radiation due to the high temperature of the melt. Thermocouples were placed at defined points in the insulation wrap, ceramic mold, and metal. Temperature in the wrap was measured with a thermocouple placed in the center of the wrap layer (marked by a blue spot in Figure 1c). Shell temperature was measured with a thermocouple placed on the leading edge of one external airfoil. Three thermocouples were used to measure the temperature in the alloy but only the results of one of them are shown because the other thermocouples failed during investment casting. The location of each thermocouple is illustrated in Figure 1c, and it was identified by nodes in the thermal model (see 'Development of the thermal model' section). The preheated assembly was placed in the vacuum casting furnace where the ceramic mold was filled by the molten metal poured at 1,549°C with a melt pouring velocity of 1,700 mm/s. The assembly was then removed from the furnace and allowed to cool. The thermal history at defined nodes of the metal, the ceramic mold, and the wrap was recorded.

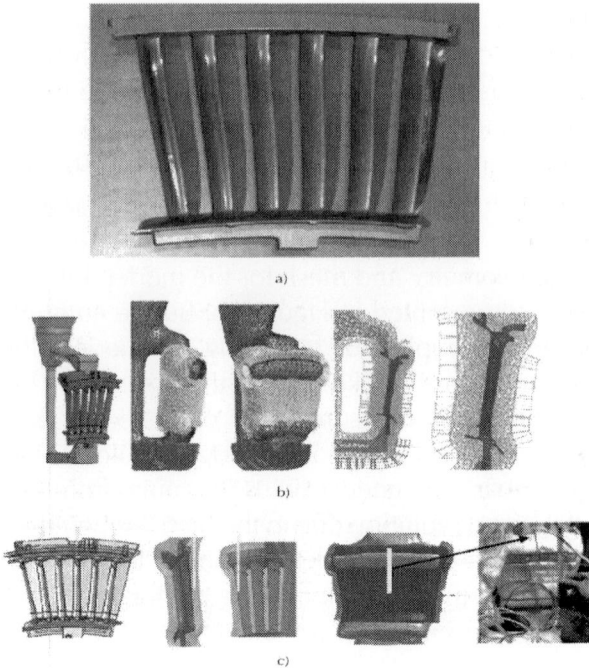

Figure 1: The NGV, model and mesh of the model for half shell, and location of thermocouples. a) The NGV produced by investment casting process; b) model and mesh of the model for half shell with insulation wrapping; c) location of thermocouples (in *yellow color*) to measure temperature on the alloy (*left in gray color*), shell (*middle in green color*), and insulation (*right in violet color*).

The as-cast NGV (Figure 1a) was cut into smaller specimens for the analysis of porosity. The selected areas for porosity evaluation are shown in the 'Porosity characterization in the as-cast new generation NGV and experimental validation of the model' section. The specimens for porosity characterization were ground and polished to the mirror-like surface using standard metallographic techniques. The optical microscope OLYMPUS BX51 was used for porosity characterization. At least three images were taken from each area of interest. Quantitative analysis of porosity (pore size and porosity volume fraction) was performed using ANALYSIS software. The pore size was given by the equivalent circle diameter due to the complex shape of pores.

Modeling

A modeling approach to investment casting of the new generation NGVs should allow the definition of the cast component, gate, mold, and insulation wrapping configuration and geometries. Starting from the component geometry, the casting process can be gradually developed and optimized, and critical design decisions can be made. Such a model has to be able to cover issues such as heat transfer (including radiation, convection, and conduction), mass transfer (mainly fluid dynamics), and phase transformations, considering at any moment the conservation of mass, momentum, and energy. And it should be able to assess the influence of the geometric and physical parameters on the porosity and structure of the as-cast NGVs. For investment casting of the new generation NGVs, most geometrical parameters such as gating, mold thickness, and wrapping scheme are already defined by the manufacturer, but most of the physical parameters remain unknown. The development of the modeling tool and definition of the thermophysical parameters are described below.

Thermal Model

Development of the Thermal Model

The basis for reliable modeling of investment casting is a very accurate prediction of the thermal history at each point of the cast. Development of the thermal model requires the optimal selection of the thermophysical parameters along with the proper establishment of boundary conditions as noted in the ASM Handbook [25]. It should also be noted that each manufacturing process has unique boundary conditions that have to be identified, understood, and characterized for the specific application being simulated. The boundary conditions can also be equipment specific, meaning that a furnace may not give rise to the same boundary conditions as another furnace under the same nominal processing conditions.

Mold filling during investment casting was modeled using the three-dimensional finite element solver ProCAST (a trademark of ESI group) by solving the conservation of mass, momentum, and heat flow equations [26]. Conservation of mass is enforced through the continuity equation

$$\frac{\delta \rho}{\delta t} + \frac{\delta(\rho u_i)}{\delta x_i} = 0$$

(1)

where u_i is the corresponding component of the velocity and ρ stands for the density. The momentum equation as used in ProCAST is given by

$$\frac{\delta u_i}{\delta t} + \rho u_j \frac{\delta u_i}{\delta x_j} + \frac{\delta}{\delta x_j}\left(p\zeta_{ij}(\mu + \mu_T)\frac{\delta u_i}{\delta u_j}\right) = \rho g_i - \frac{\mu}{Kp}u$$

(2)

where ζij is the Kronecker delta, p the pressure, g_i the gravitational acceleration, μ the viscosity, μ_T the eddy viscosity, and Kp the permeability. These equations are solved under the assumption that the spatial derivatives of viscosity are small and that the fluid is nearly incompressible.

During investment casting, heat flows by conduction through the metal, ceramic mold, and insulation wrap and is removed from the surface by natural convection and radiation. The heat flow is transient, i.e. the temperatures in the casting, mold, and insulation wrap change with time. The governing partial differential equation of heat flow by conduction is expressed as

$$\rho\frac{\delta H}{\delta T}\frac{\delta T}{\delta t} - \nabla[K\nabla T] - S(r) = 0$$

(3)

where $\nabla = \frac{\partial}{\partial x} + \frac{\partial}{\partial y} + \frac{\partial}{\partial z}$, T stands for the temperature, t for the time, K for the thermal conductivity, $S(r)$ is a spatially varying heat source, and H the enthalpy of solidification, which encompasses both the specific heat term (c_p) and the evolution of latent heat (L) during solidification according to

$$H(T) = \int_0^T c_p dr + L(1 - f_s(T))$$

(4)

where f_s is the fraction of solid.

Initial and boundary conditions for the resolution of previous equations are applied on temperature, velocity, pressure, fixed turbulent kinetic energy, fixed turbulent dissipation rate, and specific, convective, and radiation heat flux. An iterative algorithm is used to simulate solidification by solving Equation 2, finding a coherent solution between enthalpy and temperature results. Further details about this strategy can be found in [27], [28].

To solve the complex view factor radiation capability, ProCAST uses a net flux model, in which an overall energy balance for each participating surface is considered rather than tracking the reflected radiant energy from surface to surface. At a particular surface i, the radiant energy being received is denoted q_{in}, i. The outgoing flux is q_{out}, i. The net radiative heat flux is the difference of these two.

$$q_{net,i} = q_{out,i} - q_{in,i}$$

(5)

Utilizing the diffuse gray-body approximation, the outgoing radiant energy can be expressed as:

$$q_{out,i} = \sigma \, \varepsilon_i T_i^4 + (1 - \varepsilon_i) \, q_{in,i}$$

(6)

The first term of this equation represents the radiant energy which comes from direct emission. The second term is the portion of the incoming radiant energy which is being reflected by surface i.

The incoming radiant energy is a combination of the outgoing radiant energy from all participating surfaces being intercepted by surface i.

Specifically, the view factor $F_{i\text{-}j}$ is the fraction of the radiant energy leaving surface j which impinges on surface i. Thus,

$$q_{in,i} = \Sigma_{j=1}^{N} F_{i\text{-}j} \, q_{out,i}$$

(7)

where N is the total number of surfaces participating in the radiation model and the view factors are calculated from the following integral.

$$F_{i\text{-}j} = \frac{1}{A_i} \int_{Aj} \int_{Ai} \frac{\cos\theta_j \, \cos\theta_i}{\pi r^2} \, dA_i dA_j$$

(8)

where A_i is the area of surface i, θ_i the polar angle between the normal to surface i and the line between i and j, and r the magnitude of the vector between surfaces i and j.

Traditionally, Equation 8 is evaluated by numerical integration, either in the form shown or converted into an equivalent line integral. In ProCAST, the view factors are computed using a proprietary technique.

Solving Eq. 6 for $q_{in,i}$ yields

$$q_{in,i} \;=\; \left[\frac{1}{1 - \varepsilon_i}\right]\left(q_{out,i} - \sigma\, \varepsilon_i\, T_i^4\right) \tag{9}$$

Combining Equation 9 with Equation 7 gives a relationship involving the outgoing radiant fluxes only. These outgoing fluxes are known as radio sites. The final form is:

$$\sum_{j=1}^{N}\left(\zeta_{ij} - (1 - \varepsilon_i)\, F_{i\,\text{-}\,j}\right) q_{out,j} \;=\; \sigma\, \varepsilon_i\, T_i^4 \tag{10}$$

where the Kronecker delta ζ_{ij} has been included to incorporate the diagonal term. Since there are equations of the form (10), a simultaneous solution is required for a large non-symmetric system. Because of the reciprocity relation, $A_j F_{i-j} = A_j F_{j-i}$ can be transformed into a symmetric form which is more economical to solve. Multiplying (10) by

$$\frac{A_i}{1 - \varepsilon_i} \tag{11}$$

yields

$$\sum_{j=1}^{N}\left(\frac{A_i}{1 - \varepsilon_i}\, \zeta_{ij} - A_i\, F_{i\text{-}j}\right) q_{out,j} \;=\; \frac{A_i}{1 - \varepsilon_i}\, \sigma\, \varepsilon_i\, T_i^4 \tag{12}$$

which is solved for the vector of radiosities, $q_{out,i}$. The net radiant flux is obtained by combining Equation 5 and Equation 9 that gives

$$q_{net,i} \;=\; \left[\frac{\varepsilon_i}{1 - \varepsilon_i}\right]\left[\sigma\, T_i^4 - q_{out,i}\right] \tag{13}$$

This heat flux then appears as a boundary condition for the heat conduction analysis in ProCAST.

Several software packages were used to generate the thermal

model. The NGV design was created with SolidWorks software (powered by Dassault Systems SolidWorks Corporation), while the feeding system was created with Unigraphics software (powered by Unigraphics Solutions Incorporation). Both packages are linked to the casting simulation software ProCAST. Surface and volume meshes were generated by Visual-Mesh (ProCAST software package), considering a maximum distance between nodes of 2 mm inside the NGV, and 6 mm for the gating system and pouring cup. The investment casting ceramic mold was composed of layers which were created by ProCAST 3D-Mesh. The ceramic mold has an average thickness of 13 mm. The thickness of the insulation wrap was also 13 mm and was created and meshed, following the same procedure. Only one half of the mold was considered due to symmetry (Figure 1b) in order to speed up the simulations.

Data from different sources were carefully analyzed to assign the thermophysical properties to all the components of the casting system. Those sources include the ProCAST database which was described in detail by Pequet et al. [15], experimental data from industrial companies (Precicast Bilbao and Precicast Novazzano), as well as technical references from previous similar exercises. The thermophysical properties of MAR-M247 Ni-based superalloy (including temperature-dependent thermal conductivity, density, specific heat, and viscosity) were extracted from the ProCAST database (Table 2). Figure 2 illustrates dependence of these properties with temperature. The values of the liquidus and solidus temperatures (1,366°C and 1,266°C, respectively) were also taken from the ProCAST database. It should be noted that a comparison with the Lever Rule model and Scheil model (both described in the ASM Handbook [29]) was made to confirm these data.

Table 2: Thermophysical properties and boundary conditions used in the thermal model

Material	Property (units)	Value
MAR-M247	Thermal conductivity $(W \cdot m^{-1} \cdot K^{-1})$	15–35[a]
	Density $(kg \cdot m^{-3})$	7,300–8,600[a]
	Enthalpy (kJ/kg)	100–800[a]
	Viscosity $(kg \cdot m^{-1} \cdot s^{-1})$	$2\text{-}3.25 \cdot 10^{-3}$[a]
	Liquidus temperature (°C)	1,366
	Solidus temperature (°C)	1,266
Mold	Thermal conductivity $(W \cdot m^{-1} \cdot K^{-1})$	0.4–1.7[a]
	Density $(kg \cdot m^{-3})$	1,860–1,915[a]
	Specific heat $(kJ \cdot kg^{-1} \cdot K^{-1})$	0.7–1.3[a]
	Emissivity	0.7
Insulation wrap	Thermal conductivity $(W \cdot m^{-1} \cdot K^{-1})$	0.1–0.5[a]
	Specific heat $(kJ \cdot kg^{-1} \cdot K^{-1})$	0.9–1.3[a]
	Emissivity	0.7
Metal mold	HTC $(W \cdot m^{-2} \cdot K^{-1})$	200–2,500[a]
Mold wrap	HTC $(W \cdot m^{-2} \cdot K^{-1})$	100
Mold enclosure	HTC $(W \cdot m^{-2} \cdot K^{-1})$	3
Wrap enclosure	HTC $(W \cdot m^{-2} \cdot K^{-1})$	10.6
Enclosure	Emissivity	0.9
Others	*Units*	*Value*
Melt pouring velocity	(mm/s)	1,700
Melt temperature	(°C)	1,549
Preheating temperature	(°C)	1,200

[a]The value depends on the temperature.

Torroba *et al.*

Torroba *et al. Integrating Materials and Manufacturing Innovation*
2014 **3**:25   doi:10.1186/s40192-014-0025-5

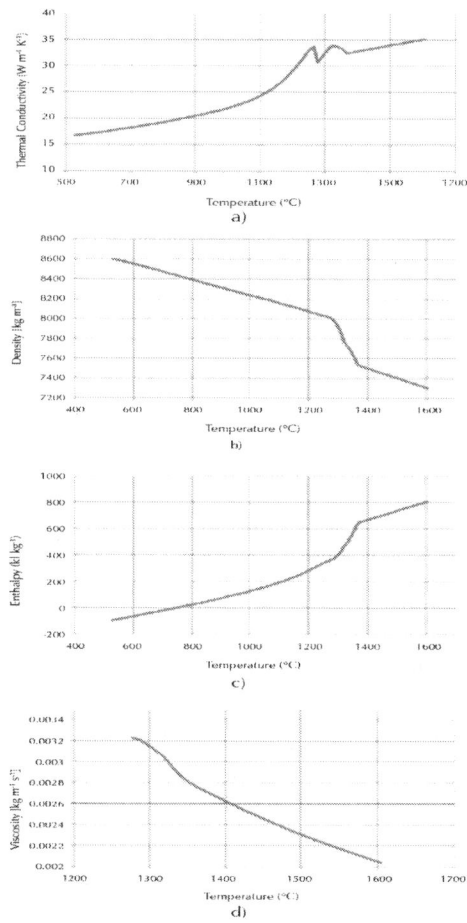

Figure 2: Properties of MAR-M247 Ni-based superalloy vs. temperature. a)
Thermal conductivity, b) density, c) enthalpy, and d)viscosity.

Regarding the ceramic mold and wrap insulation, the density
and specific heat as functions of temperature were taken from the
ProCAST database (Table 2). Values for the thermal conductivity as
a function of temperature (Figures 3a and 4a) were obtained by an
inverse simulation procedure by comparing the simulation results for

simple casting geometries with experimental data generated earlier by Precicast Bilbao. The description of the inverse simulation procedure can be found in O'Mahoney and Browne [30]. Figures 3 and 4 illustrate the variation of these properties with temperature for the ceramic mold and wrap, respectively.

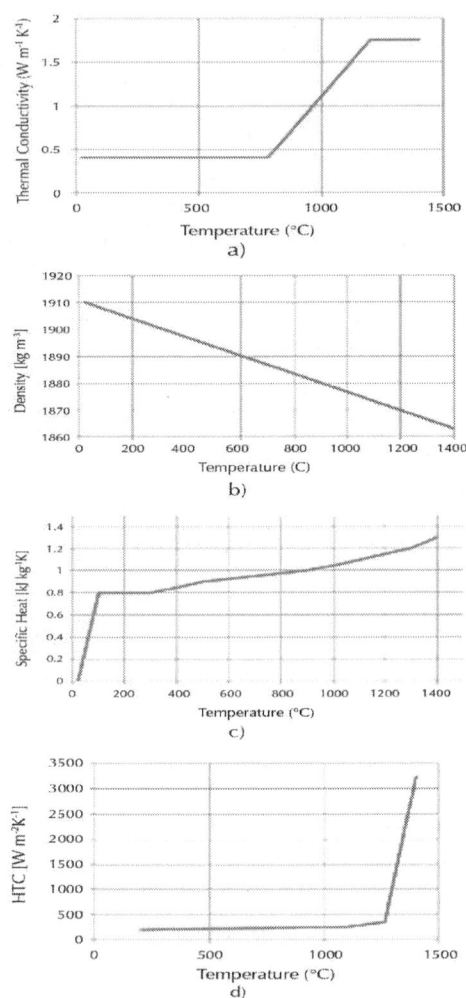

Figure 3: Properties of ceramic mold vs. temperature. a) Thermal conductivity, b) density, c) specific heat, and d) heat transfer coefficient (HTC) at the metal-mold interface.

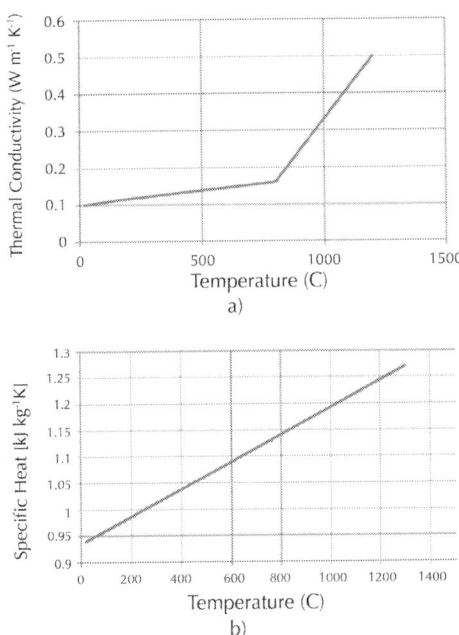

Figure 4: Properties of wrap vs. temperature. a) Thermal conductivity and b) specific heat.

It is known that pouring of the melt at high temperature leads to radiation heat loss. As this heat loss is not always correctly taken into account during the modeling process, the value of mold conductivity at high temperatures should be increased to account for this phenomenon. Experimental studies on this topic were earlier carried out by Precicast Bilbao and the experimental data from earlier measurements using the laser flash method (according to the ASTM E1461-07 standard) were considered. Analysis of all available data led to a final interval of mold conductivity in the range from 0.4 to 1.75 $W \cdot m^{-1} \cdot K^{-1}$ (Table 2), which is in a very good accordance with the data provided by Konrad et al. [31] for low temperatures, and coincide with the experimental data measured by the laser flash method at high temperature (Figure 3a). The ProCAST database, data from the manufacturer of the kaolin wool, and Precicast Bilbao were considered to define the thermal conductivity of the insulation wrap. The final values of the thermal conductivity in the insulation wrap were in the range of 0.1 to 0.5 $W \cdot m^{-1} \cdot K^{-1}$ (Figure 4a and Table 2).

A suitable temperature-dependent functional form (shown in Figure 3d) was used to determine the values of heat transfer coefficient (HTC) at the metal-mold interface. It is known that the HTC at the metal-mold interface is influenced by many factors such as casting geometry, pouring and preheating temperature, mold thickness, etc. Inverse and direct simulations were carried to obtain the final form of this function, which is plotted in Figure 3d as a function of temperature. This function is slightly different from the one proposed by the ProCAST database as was demonstrated by Santos et al. [32] and Dong et al. [33]. Nevertheless, the final HTC at the metal-mold interface was in a very good agreement with the data reported in the literature for molten Ni-based superalloys in contact with ceramic molds. For example, Sahai and Overfelt [34] reported a HTC in the range 50–5,000 W m^{-2} K^{-1} for IN-718 Ni-based superalloy. The HTC at the mold-wrap, mold-enclosure, and wrap-enclosure interfaces have less influence on the final result of the thermal model as shown by Yuang et al. [35]. Thus, it was assumed that they were constant with temperature and time, and the data from the ProCAST database were used (Table 2). Values of emissivity for mold, wrap, and enclosure were also taken from the ProCAST database, and the environmental conditions were fitted with those registered during experimental casting trials (Table 2). The pouring of the melt into the mold was introduced in the model by the definition of a planar surface on the top of the pouring cap, where a velocity to the liquid was applied to simulate the pouring process. The preheating temperature of the mold and temperature of melt poured into the mold were also specified (Table 2). The same filling steps performed during the experimental procedure were simulated by the software, using 2 s of filling time to introduce the molten alloy into the mold. Thus, the full solidification process was completed 830 s after the pouring. All the experimental data were taken into account during the simulations to synchronize the experimental data with the simulation results. The simulation process was operated by the ProCAST Parallel Solver with four processors (2.40 GHz) and took nearly 11 h to simulate the whole thermal history of the NGV investment casting process.

Experimental Validation of the Thermal Model

Experimental casting trials were carried out for validation of the thermal model as described above. Figure 5 illustrates the experimental temperature-time plots for metal, ceramic mold, and insulation wrap during investment casting. Temperature recording was started once the thermocouples were located in the defined spots. The thermocouple placed in the metal is close to reach the preheating temperature 1,200°C, while the thermocouple placed in the mold records a temperature slightly over 1,100°C. In the readings from the thermocouple fixed to the insulation wrap, temperature rises up to 900°C. Significant difference of temperatures between metal and wrap was registered at the beginning since it took time to place correctly each thermocouple into its location. This loss of time leads to partial cooling of the mold that, in turn, increases the temperature gradient between metal and mold.

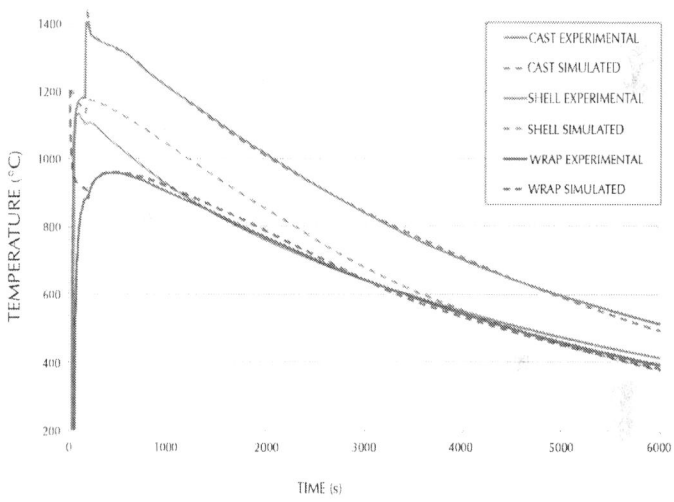

Figure 5: Temperature-time plots for metal, mold (shell), and insulation wrap. Comparison of the thermal model prediction with the experimental results.

After the mold entered the vacuum casting furnace, vacuum was pumped and melt was poured into the ceramic mold. The thermocouples placed in the metal and wrap clearly registered this event by showing a rapid temperature rise, whereas the thermocouple located in the ceramic mold showed a temperature decrease (Figure 5). The liquidus-solidus transition in the metal can be easily identified in the experimental temperature-time plot because of the reduced slope. The cooling rate increases once the melt is solidified.

The temperature-time plots generated by the thermal model are compared with the experimental results in Figure 5. A very good agreement is observed for the thermal history in the metal and in the wrap, where the simulation results match very well the experimental results during first 6,000 s of the solidification/cooling process. However, a difference of nearly 100°C is found between the predicted temperature and the experimental data in the ceramic mold. Despite the close location of the thermocouple to the inner surface of the ceramic mold, the temperature registered by this thermocouple hardly achieves 1,100°C, though the melt was poured into the ceramic mold at 1,459°C. The reasons for such discrepancy are discussed in the 'Accuracy of the thermal model' section.

A proper prediction of the liquidus-solidus transition has to be achieved in a reliable thermal model. A deeper analysis of the liquidus-solidus transition is found in Figure 6, which shows a perfect match between simulation and experimental results. The most significant deviation between numerical predictions and experimental results occurs at 300 s after pouring, and the difference is just 4°C.

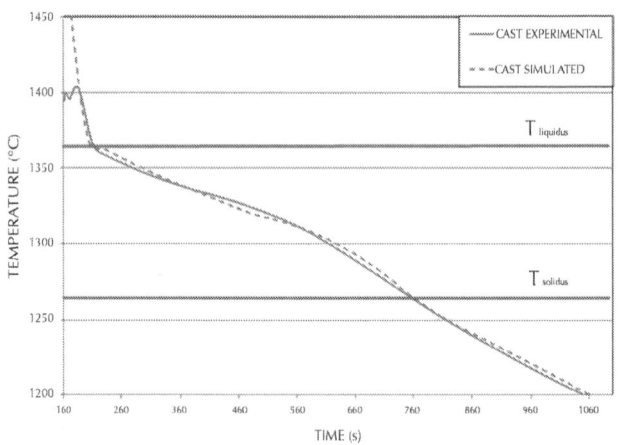

Figure 6: Comparison of experimental measurements with model prediction. Comparison of experimental measurements from metal with the thermal model prediction in the temperature range between liquidus (1,366°C) and solidus (1,266°C).

Since the solidification process of the metal and its thermal history are correctly described, the thermal model can be further utilized to predict the microstructure and defects of the as-cast parts. The next section of this manuscript focuses on the ProCAST model for porosity prediction, which is developed on the basis of the thermal model.

Model for Porosity Prediction

Description of the Model

The ProCAST tool was employed in this work to simulate the development of porosity during investment casting. The physical basis of the model is following. The key variable is the fraction of solid (FS) which extends from FS =0 for liquid to FS =1 for solid, as shown schematically in Figure 7. When the melt solidifies, pockets of liquid are created, surrounded by a mushy zone and then a solid shell. Automatically, the casting is divided into 'regions' having the FS <1. These 'regions' are bounded by iso surfaces. As solidification proceeds and depending upon the complexity of the geometry, the number of 'regions' may increase with time, i.e. one 'region' can be split in more 'regions'. The 'region' disappears once all nodes are completely solidified.

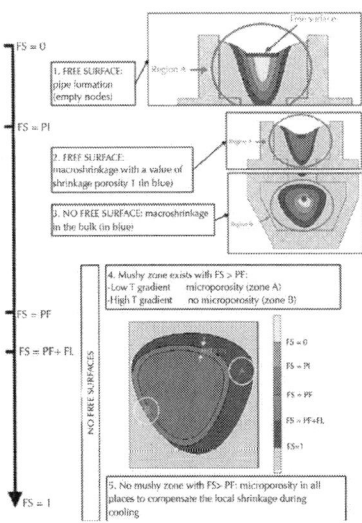

Figure 7: Schematic presentation of the porosity prediction model.

The model is based on comparison of the local FS with a few parameters describing 'key stages' of solidification which determine the porosity in the cast. The first parameter, PI, is a measure of limit of local solid fraction under the surface until piping[a] on the surface can occur. In other words, the model predicts formation of pipe (empty nodes) while FS < PI (Figure 7). In the present calculations, the default value recommended in the ProCAST database PI =0.3 was used, i.e. piping occurs until the solid fraction reaches 30%. No porosity formation takes place in the bulk of the casting while local FS < PI. The second important parameter, PF, is a limit of solid fraction until the liquid can still feed a hot area. PF =0.7 (the default value recommended in the ProCAST database) was used in the present simulations. The model predicts the formation of pipe in the form of a shrinkage pore on the surface while PI < FS < PF (Figure 7). If there are no nodes of the 'region' on the free surface having PI < FS < PF, no pipe can be formed and the model predicts macroshrinkage in the bulk of casting (Figure 7). In this case, the macropore nucleates and grows at the highest point of the liquid pocket.

[a]Piping is the formation of pipes during solidification. Pipes are open-air shrinkage defects which form at the surface of the casting and burrow into the casting.

According to the model, microporosity can appear only in the zone having PF < FS <1. The third parameter FL, critical feeding length, is introduced into the model to predict microporosity. The FL value depends upon the size of the mushy zone and thus, the size of the casting. In the present calculations, FL =0.005, following the value recommended in the ProCAST database. Two scenarios are possible in the bulk of casting:

- There is still some mushy zone (liquid) below PF. Microporosity forms only at the distance higher than FL from the PF isosurface (see zone A for the corresponding situation in Figure 7). The amount of microporosity is equal to the density change between the local solid fraction and 1. No micropores can form in the case of high-temperature gradients, since the distance between PF and solidus isosurface is smaller than FL (see zone B for the corresponding situation in Figure 7). On the contrary, low-temperature gradient promotes formation of microporosity (see zone A for the corresponding situation in Figure 7).

- There is no more mushy zone in zones with PF < FS <1. In this case, the parameter FL is not active. In this case, there can be microporosity in the whole region with PF < FS <1 to compensate the shrinkage during cooling. The level of porosity is calculated based on the change of the density for each node as solidification takes place. This variation in the density of the material allows the software to compute the volume corresponding to shrinkage porosity as the limit value of PF is achieved in the nodes.

The modeling results are displayed in ViewCast software. The unit is volume fraction [%].

The results are classified as follows:

the porosity values below 1% correspond to microporosity;

the porosity values in the range between 1% and 2.3% correspond to shrinkage porosity;

the porosity values above 2.3% correspond to the macroporosity.

The present tool was applied for porosity prediction in the as-cast NGVs, and the simulation outcomes are presented in the next section.

RESULTS AND DISCUSSION

Porosity Characterization in the As-cast New Generation NGV and Experimental Validation of the Model

Figure 8 shows the porosity predictions (left) and the experimental data (right) of transversal section of a solid vane in the as-cast NGV. The optical micrographs corresponding to the trailing edge (zone a), middle part (zone b) and leading edge (zone c) are also plotted in Figure 8. The analysis of these results shows a good agreement between simulation predictions and experimental results in all zones. The highest level of porosity (≤2.91%) is predicted for the middle part of the solid vane and it is in quantitative agreement with the experimental shrinkage porosity of 3.07% in this area (Figure 8b and Table 3). The average pore size is 22 µm. Analysis of the histogram of pore size distribution shows that the frequency of pores decreases with increasing size and a few macropores with a size up to 196 µm are present in the middle part (Figure 9a). For the leading edge, lower levels of shrinkage porosity (≤2.17%) are predicted by the model, but this prediction overestimates the experimental result, 0.63% (Figure 8a and Table 3). The average pore size slightly decreases to 20.4 µm and the maximum pore size does not exceed 79 µm (Figure 9b). Finally, shrinkage porosity was not predicted in the trailing edge and this is confirmed by experimental study (Figure 8a and Table 3).

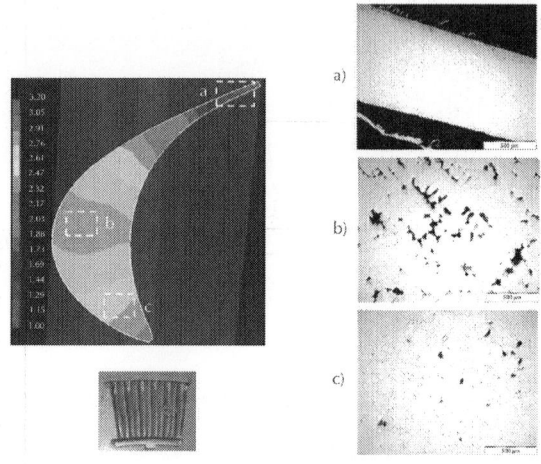

Figure 8: Porosity prediction by the ProCAST model and optical microscopy images of porosity. Porosity prediction by the ProCAST model (*left*) and optical microscopy images of porosity for the defined areas (*right*). The location of the analyzed section in the NGV is marked by *red circle*.

Table 3: The porosity characteristics of the transversal section of the solid vane on Figure8

	Leading edge	Middle part	Trailing edge
Local porosity fraction [%]	0.63	3.07	No porosity
Average pore size [µm]	20.4	22.0	-

Torroba *et al.*

Torroba *et al. Integrating Materials and Manufacturing Innovation* 2014 **3**:25   doi:10.1186/s40192-014-0025-5

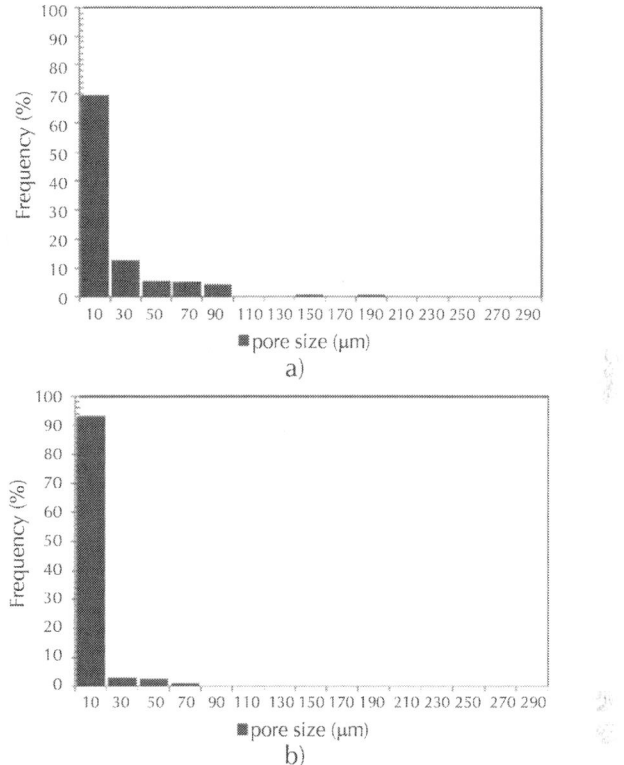

Figure 9: Histograms of pore size distribution on transversal section of the solid vane. a) The middle part (Figure 8b) and b) the leading edge (Figure 8c).

Figure 10 illustrates the outcomes of porosity modeling for the longitudinal section of the solid vane along with the optical microscopy images for selected areas. A good correlation between simulation and experimental results is observed. The simulation results show the highest level of macroporosity ($\leq$2.90%) in the red zone (b). The experimental evaluation of porosity in the red zone (b) yields porosity of 4.87% with the average pore size of 113 μm (Table 4). The frequency of pores decreases with increasing size and a few macropores with a size up to 280 μm are present in the hot spot (Figure 11b). Macroporosity of

≤2.83% is expected in the zone (c) according to the model, whereas the experimental results show macroporosity of 4.82% (Table 4). The amount of the large pores decreases in this area (Figure 11c) and the average pore size is reduced to 27 μm, correspondingly (Table 4). A decrease of porosity should take place in zones a and d and shrinkage porosity of ≤2.43% is predicted, though the experimental characterization of these areas shows much lower values of porosity (Table 4).

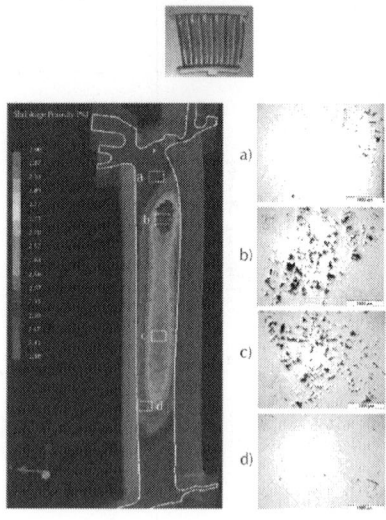

Figure 10: Porosity prediction by the ProCAST and optical microscopy images of defined areas of the solid vane. Porosity prediction by the ProCAST model (*left*) and optical microscopy images of porosity for defined areas (*right*) of the solid vane (marked by *red dashed line* on the NGV icon).

Table 4: The porosity characteristics of the longitudinal section of the solid vane on Figure10

Area	a	b	c	d
Local porosity fraction [%]	0.19	4.87	4.82	0.14
Average pore size [μm]	32	113	27	29

Torroba *et al.*

Torroba *et al. Integrating Materials and Manufacturing Innovation* 2014 **3**:25   doi:10.1186/s40192-014-0025-5

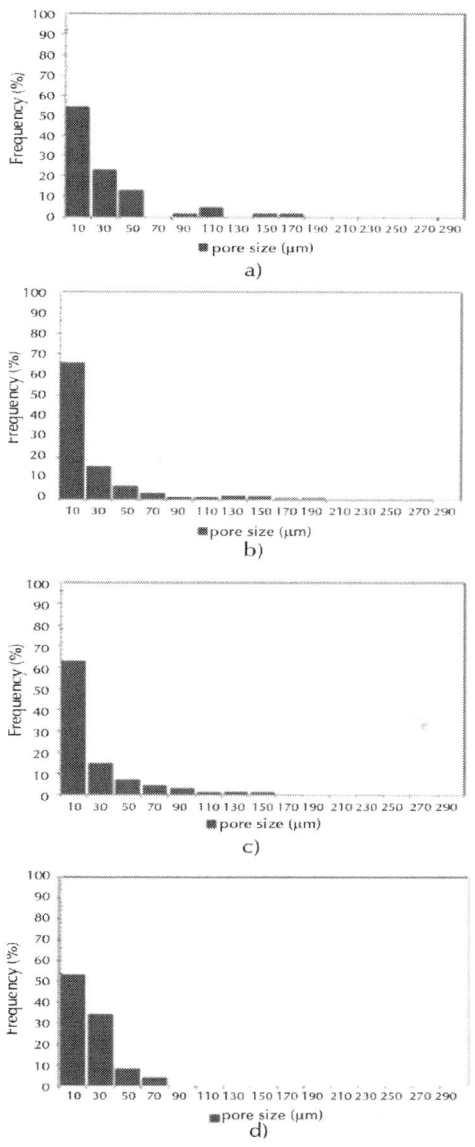

Figure 11: Histograms of pore size distribution on the longitudinal section of the solid vane. a) Area on Figure 10a, b) area on Figure 10b,c) area on Figure 10c, and d) area on Figure 10d.

The simulation results were also validated for the hollow vane of the new generation NGV (Figure 12). Generally lower porosity is predicted at the top (≤2.17%) and bottom (≤2.47%) sections of the hollow vane since the small thickness of the walls hastens the solidification process, thus reducing porosity. The highest porosity (≤2.73%) is expected in the midsection of the hollow vane as it solidifies last. The experimental data of porosity in all these areas follow the trends predicted by the simulation (Figure 12), although the modeling results tend to overestimate the porosity in the hollow vane (Table 5).

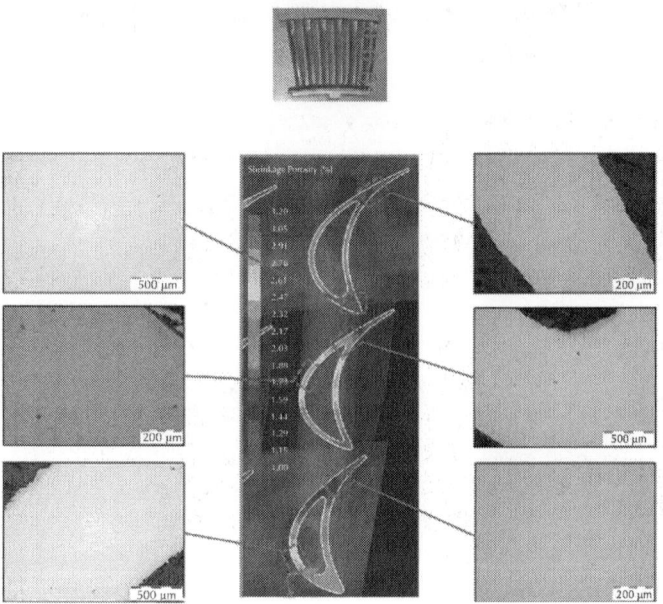

Figure 12: Porosity results from simulation (*middle*) and experimental analysis (*left* and *right hand sides*). Top, middle, and bottom sections of the hollow vane are considered. The locations of the analyzed section in the NGV are marked by *red circles*.

Table 5: The porosity characteristics of the transversal section of the hollow vane on Figure12

Area	Top	Middle	Bottom
Average porosity fraction [%]	0.06	0.08	0.08
Average pore size [µm]	19	17	15

Torroba et al.

Torroba et al. Integrating Materials and Manufacturing Innovation 2014 **3**:25 doi:10.1186/s40192-014-0025-5

Accuracy of the Thermal Model

The analysis of the simulation results and their comparison with the experimental data clearly show that the thermal history of the metal and wrap is very well described by the thermal model during the solidification and cooling processes (Figure 5). The differences between predictions and experimental measurements in the mold do not seem to be due to the model. The thermal plot measured from the ceramic mold seems to underestimate its real thermal history since the thermocouples record slightly higher temperatures for the insulation wrap compared to the ceramic mold in the time range of 1,300–3,000 s (Figure 5), which cannot be true. This discrepancy can be rationalized on the basis of the shortcomings of the experimental procedure utilized to measure the temperature in the ceramic mold. In particular,

- Cement was used to fix the thermocouple to the ceramic mold;
- Some 'air gaps' can appear between the thermocouple and cement due to significant thermal expansion/contraction;
- There could also be some deviations from the correct positioning of the thermocouple in the ceramic mold during its fixing to the ceramic mold, since this operation has to be performed manually at extreme conditions in limited time.

The 'air gaps' and cement can significantly reduce the heat transfer from the mold to the thermocouple since they have lower thermal conductivity compared to the ceramic mold. Therefore, the experimental measurements on the mold can yield lower temperatures than the real temperatures, as seen from Figure 5. These shortcomings could also lead to the drop of temperature readings from the thermocouple placed in the ceramic mold at the moment of melt pouring which could result in thermocouple shifting due to thermal expansion of ceramic mold (Figure 5).

It should be noted that the possible 'air gaps' and the cement were not considered in the thermal model, since it would increase enormously the time required for calculations. Another experimental procedure should be developed for more accurate recording of thermal history in the ceramic mold. Nevertheless, the thermal histories of metal and wrap were accurately predicted as a result of the right selection of thermophysical parameters of the ceramic mold in the thermal model.

Accuracy of the Model for Porosity Prediction

Analysis of the overall porosity prediction shows that the model predicts well the location of hot spots and areas prone to porosity formation throughout the NGV. These areas are located mainly in the solid vanes and this was confirmed by the experimental characterization of the as-cast NGV. A very good match between the simulation predictions and the experimental results was found in many NGV areas. However, the model tends to slightly underestimate porosity in the areas located in the thickest parts of the NGV (Figure 10b). This discrepancy could be related to liquid pools which can be formed in those areas during solidification, as reported recently by Kang et al. [22]. Another shortcoming of the model is the overestimation of shrinkage porosity in the thinnest parts of the NGV, which are the hollow vanes with a wall thickness nearly 1 mm. This effect could be explained by formation of skin which can significantly affect the local thermal history of the metal in the thin parts. It should be noted that the rapid skin formation due to freezing of melt with a colder ceramic mold is not taken into account by the model.

All in all, it can be outlined that the simulation tool for porosity prediction can be successfully utilized for further improvement of NGV design. Its application can significantly reduce the number of expensive experimental casting trials which are typically required to find the suitable casting parameters and to develop a manufacturing route for investment casting of complex shape parts at industrial scale.

CONCLUSIONS

Investment casting of NGV from Ni-based superalloys was simulated by means of a finite element model. The simulation strategy is targeted to predict the heat exchange during solidification and cooling and the porosity. The casting assembly, consisting of the hollow ceramic mold with NGV-shape interior and insulation wrap, is created and meshed. The thermophysical parameters and boundary conditions are defined for all the parts of the casting assembly, and simulation is carried out using ProCAST. Experimental casting trials are performed for validation of the developed models.

The thermal history of the metal and the insulation wrap during investment casting was accurately predicted. The critical thermal-physical parameters of the casting system were obtained either from the literature or by an inverse simulation procedure by comparing the simulation results for simple casting geometries with experimental data. The hot spots and areas with enhanced porosity which are located in the thickest parts of the NGV were accurately predicted. In addition, the porosity predictions were in good agreement with the experimental results in many NGV areas. The shortcomings of the porosity predictions include a slight underestimation of porosity in some very thick areas and an overestimation of shrinkage porosity in the thinnest parts of the NGV. It is concluded that the developed modeling tool can be successfully utilized for further improvement of NGV design, allowing to minimize the number of casting trials.

AUTHORS' CONTRIBUTIONS

AJT applied thermal model and porosity prediction model and prepared the initial draft of the manuscript. OK and LC provided materials science guidance and expertise in modeling. LM performed experimental casting trials. ECM contributed with the overall development of the main concepts presented in this paper. MR performed quantitative analysis of porosity in the as-cast NGV. SM and IS helped with the validation of the porosity tool and manuscript writing. JL formulated the concept of this work and conceived the workflow, as well as provided materials science guidance. All authors contributed to the manuscript. The final version was prepared by IS and JL and approved by all authors. All authors read and approved the final manuscript.

ACKNOWLEDGEMENTS

This investigation was carried out in frame of the VANCAST project (EU, FP7, ERA-NET MATERA+). SM and IS acknowledge gratefully the Spanish Ministry of Economy and Competitiveness for financial support through the Ramon y Cajal fellowships. Prof. A. Zryd (Maxwell Technologies SA) and Dr. A. Faes (CSEM SA) are greatly acknowledged for the inverse simulations of experimental casting trials of easy geometry parts as those results constituted the seed for experimental work which had led to this manuscript.

REFERENCES

1.   Razak AMY (2007) Industrial gas turbines: performance and operability. Woodhead Publishing Limited, Cambridge, UK.

2.   Reed RC (2006) The Superalloys: Fundamentals and Applications. Cambridge University Press, Cambridge, UK.

3.   Pattnaik S, Karunakar DB, Jha PK (2012) Developments in investment casting process—a review. J Mater Proc Tech 212:2332-2348 doi:10.1016/j.jmatprotec.2012.06.003

4.   Anglada E, Meléndez A, Maestro L, Domiguez I (2013) Adjustment of numerical simulation model to the investment casting process. Proc Eng 63:75-83 doi:10.1016/j.proeng.2013.08.272

5.   Rafique MMA, Iqbal J (2009) Modeling and simulation of heat transfer phenomena during investment casting. Int J Heat Mass Transf 52:2132-2139 doi:10.1016/j.ijheatmasstransfer.2008.11.007

6.   Stefanescu DM (2009) Science and Engineering of Casting Solidification. Springer Science + Business Media, New York, NY, USA.

7.   Piwonka TS, Flemings MC (1966) Pore formation in solidification. Trans AIME 236(8):1157-1165

8.   Pellini WS (1953) Factors which determine riser adequacy and feeding range. AFS Transactions 61:61-80

9. Niyama E, Uchida T, Morikawa M, Saito S (1981) Predicting shrinkage in large steel castings from temperature gradient calculations. AFS Int Cast Met J 6(2):16-22

10. Carlson KD, Beckermann C (2009) Prediction of shrinkage pore volume fraction using a dimensionless Niyama criterion. Metall Mater Trans A 40:163-175 doi:10.1007/s11661-008-9715-y

11. Kubo K, Pehlke RD (1985) Mathematical modeling of porosity formation in solidification. Metall Mater Trans B 16:359-366 doi: 10.1007/BF02679728

12. Lee PD, Hunt JD (2001) Hydrogen porosity in directionally solidified aluminium copper alloys: a mathematical model. Acta Mater 49:1383-1398 doi:10.1016/S1359-6454(01)00043-X

13. Lee PD, Chirazi A, Atwood RC, Wang W (2004) Multiscale modeling of solidification microstructures, including microsegregation and microporosity, in an Al-Si-Cu alloy. Mater Sci Eng A 365:57-65 doi:10.1016/j.msea.2003.09.007

14. Carlson KD, Lin Z, Beckermann C (2007) Modeling the effect of finite-rate hydrogen diffusion on porosity formation in aluminum alloys. Metall Mater Trans B 38:541-555 doi:10.1007/s11663-006-9013-2

15. Pequet C, Rappaz M, Gremaud M (2002) Modeling of microporosity, macroporosity, and pipe-shrinkage formation during the solidification of alloys using a mushy-zone refinement method: applications to aluminum alloys. Metall Mater Trans A 33:2095-2106 doi:10.1007/s11661-002-0041-5

16. Couturier G, Rappaz M (2006) Effect of volatile elements on porosity formation in solidifying alloys. Model Simul Mater Sci Eng 14(2):253-271 doi:10.1088/0965-0393/14/2/009

17. Couturier G, Rappaz M (2006) Modeling of porosity formation in multicomponent alloys in the presence of several dissolved gases and volatile solute elements. TMS Annual Meeting, San Antonio, TX, USA.

18. Stefanescu DM (2005) Computer simulation of shrinkage related defects in metal castings—a review. Inter J Cast Metal Res 18(3):129-143

19. Lee PD, Chirazi A, See D (2001) Modeling microporosity in aluminum–silicon alloys: a review. J Light Metals 1:15-30 doi:10.1016/S1471-5317(00)00003-1

20. Overfelt RA, Sahai V, Ko YK, Berry JT (1994) Porosity in cast equiaxed alloy 718. In: Loria EA (ed) Proceedings of the TMS Meeting, p189

21. Monastyrskiy VP (2010) Modeling of porosity formation in Ni-based superalloys. In: Choi JK (ed) Proceedings of the 8th Pacific Rim International Conference on Modeling of Casting and Solidification Process, p89

22. Kang M, Gao H, Wang J, Ling L, Sun B (2013) Prediction of microporosity in complex thin-wall castings with the dimensionless Niyama criterion. Materials 6:1789-1802

23. Calba L, Lefebvre D (2008) Modeling the investment casting process. ESI-GROUP Resource Center, Paris.

24. Harris K, Erickson GL, Schwer RE (1984) MAR-M247 derivations— CM247 LC DS alloy, CMSX single crystal alloys, properties and performance. In: Gell M, Kortovich CS, Bricknell RH, Kent WB, Radvich JF (eds) Proceedings of the 5th International Symposium on Superalloys, TMS, p221

25. Handbook ASM (2010) Metals Process Simulation. ASM International, Ohio, USA.

26. (2007) Version 6.1. ESI software, France.

27. Rappaz M, Bellet M, Deville M, Snyder R (2002) Numerical modeling in materials science and engineering. Springer-Verlag, Berlin, Germany.

28. Dantzig JA, Rappaz M (2009) Solidification. EPFL-Press, Lausanne, Switzerland.

29. Handbook ASM (2008) Casting. ASM International, Ohio, USA.

30. O'Mahoney D, Browne DJ (2000) Use of experiment and an inverse method to study interface heat transfer during solidification in the investment casting process. Exper Thermal Fluid Sci 22:111-122 doi:10.1016/S0894-1777(00)00014-5

31. Konrad CH, Brunner M, Kyrgyzbaev K, Völkl R, Glatzel U (2011) Determination of heat transfer coefficient and ceramic mold material parameters for alloy IN738LC investment castings. J Mater Proc Tech 211:181-18 doi:10.1016/j.jmatprotec.2010.08.031

32. Santos CA, Quaresma JMV, Garcia A (2001) Determination of transient interfacial heat transfer coefficients in chill mold castings. J Alloys Compd 319:174-186 doi: 10.1016/S0925-8388(01)00904-5

33. Dong Y, Bu K, Dou Y, Zhang D (2011) Determination of interfacial heat-transfer coefficient during investment-casting process of single-crystal blades. J Mater Proc Tech 211:2123-2131 doi:10.1016/j.jmatprotec.2011.07.012

34. Sahai V, Overfelt RA (1995) Contact conductance simulation for alloy 718 investment casting of various geometries. Tran Amer F 103:627-632

35. Yuang XL, Lee PD, Brooks RF, Wunderlich R (2004) The sensitivity of investment casting simulations to the accuracy of thermophysical properties values. Proceedings of the International Symposium on Superalloys, TMS. 951

A High-Resolution Resistive Probe for Nonlinear Analysis of Two-Phase Flows

L. Cantelli, A. Fichera, and A. Pagano

DIIM, Università Degli Studi di Catania, Viale A. Doria n. 6, 95125 Catania, Italy

ABSTRACT

Two-phase flow dynamics are highly complex, due to the strong coupling of various independent mechanisms and as demonstrated by the existence of a variety of flow patterns. The adoption of appropriate tools for nonlinear time series analysis tools may lead to a deeper insight in this complexity but requires high quality time series. This study describes a procedure appositely assessed in order to realize an impedance void fraction sensor of resistive type characterized by high-spatial and -temporal resolution. These characteristics have been accomplished through an appropriate geometrical design of the probe electrodes, aiming at obtaining a thin measurement volume so to improve the probe spatial resolution, and through the electronic assessment of the data acquisition system, improving its temporal resolution. A new calibration procedure has been also defined, based on an estimation of void fraction through a code for automatic extraction of bubble contours and the correction of image distortions.

INTRODUCTION

Gas-liquid two-phase flows in pipes are at the basis of a wide variety of heat and mass transfer applications, ranging from power generation to chemical, processing, and oil plants. This kind of flows is characterized by high complexity, as a consequence of the strong coupling of several mechanisms and of the dependence on various factors, the most important of which are the differential action of gravity on the two-phases and the effect of shear and surface tension forces at their interface. In particular, depending on the hydrodynamic equilibrium at the interface between the two phases, several different flow patterns can be identified, each of which can be characterized in terms of the peculiar distribution of the two phases and of pressure drop. The type of flow pattern represents one of the fundamental factors governing the dynamics of two phase systems; therefore, flow patterns identification and classification are of primary importance for both engineering design purposes and for the monitoring of applications involving two-phase flows.

Though several different names and classifications have been proposed by researchers, some flow patterns have been recognized as typical and reported in several classifications, such as works in [1–5]. For the case of vertical pipes considered in the present study, the basic flow pattern classification distinguishes between bubbly, slug, churn, and annular flow.

In the bubbly flow small diameter gas bubbles are dispersed in the liquid phase. The main characteristic is that coalescence phenomena, though present, are unable to produce gas bubbles occupying the pipe section, as it happens in the slug flow.

The slug flow consists in an intermittent flow of Taylor bubbles alternated to liquid slugs. Moreover, the liquid slugs can be aerated or not, as a consequence of the entrainment of small gas bubbles. The respective flow rates of the two phases are determinant for the development of the bubble, which may range from very short to elongated, depending on the relative importance of the three main parts in which the bubble can be subdivided. These are:

- the head region, where the fraction of the pipe section occupied by gas rapidly grows from zero to approximately the whole section, that is, except for the thin liquid film separating the gas from the tube wall;
- the central region, where the pipe section is mainly occupied by the gas phase, again except for the liquid film at the wall, which is often subjected to longitudinal oscillations, especially if the central region is sufficiently developed;
- the tail region, where the gas occupying the pipe section abruptly falls to approximately zero, that is, except for the shedding of relatively small bubbles in the liquid slug that follows the Taylor bubble.

Depending on the development of the central region of the bubble, it is possible to draw a main distinction in the class of slug flow. In particular, cap flow occurs when the head and tail regions of the bubble are consecutive and no central region can be distinguished. Cap flows usually occur at relatively low values of the gas mass flow rate; therefore, the cap-shaped bubbles are usually separated by long liquid slugs. In the proper slug flow, the axial development of the head, central, and tail regions of the bubble is comparable, whereas the plug flow is characterized by elongated Taylor bubbles, that is, it occurs when the extension of the central region is markedly predominant with respect to the head and tail regions. The distinction between slug and plug flow remains somewhat controversial; nonetheless, plug flow occurs at relatively higher values of the gas mass flow rate and, therefore, the liquid slugs separating the gas plugs are usually shorter and more aerated than those separating the gas bubbles in slug flow.

The churn flow is a highly turbulent flow that occurs for higher values of the gas mass flow rate. In this flow pattern, the waving liquid film still drains down the wall, as in elongated Taylor bubbles, but occasionally bridges the tube wall forming short and highly aerated slugs. Finally, the annular flow consists of a thin annular film of liquid on the tube wall on which small ripples, occasionally interspersed with large disturbance waves, flow in a regular manner along the tube wall. These flow patterns occur when the gas mass flow rate is high enough to guarantee the stability of the liquid-gas interface.

The characterization of the various flow patterns can be performed on the basis of the analysis of void fraction-related signal fluctuations. In fact, under normal gravity conditions, the void fraction represents the most important variable for the description of the distribution of the two-phases within the pipe section, and is therefore a fundamental parameter for the identification and classification of the flow regime. The accuracy of the classification strategy strongly depends on the spatial and temporal resolution of the technique adopted for the measure of the void fraction. Several of these techniques have been proposed; the most common are based on the measure of the electrical impedance of two phase mixtures [2, 6, 7], of the optical scattering of the interface between the two phases [8], or of the pressure difference observed along a specified piece of the pipe [9].

The design of optical sensors allow to perform local measurement but is severely affected by disturbances caused by the nonlinearity of the scattering on curved interfaces; therefore, in general, the quality of their experimental time series is very poor. In contrast, differential pressure sensors are not sensitive to phenomena induced by the curvature of the interface between the two phases but cannot be designed in order to obtain a reliable local measurement, as they require the two pressure tips to be placed at a sufficient distance along the pipe axis. Moreover, the structure of the time series of neither optical nor differential pressure sensors can be directly related to the physical distribution of the two phase within the flow pattern.

In contrast to the previous studies, impedance measurements seem to be particularly reliable [8]. They are non-intrusive and, most important, less dependent both on internal disturbances and external factors. Two main classes of impedance sensors have been proposed in the literature: resistive sensors [2, 6, 10, and 11] and capacitance sensors [7, 9]. This kind of sensors presents some relevant advantages with respect to the previous. In particular, a strong correspondence exists between the shape of the oscillations of the experimental time series and the physical distribution of the two phase within the pipe section for the various flow patterns; this is remarkably verified for flow patterns characterized by pulsations of a relatively stable nature, such as, for example, for cap, slug, and plug flows, but also for annular flows with a wavy liquid film at the pipe wall. On the basis of this correspondence, for example, the identification of the passage of a bubble through the measurement volume, as well as its shape and length, is clearly reported in the time series. A second important advantage, partly depending on the previous, is the minor influence of secondary noisy phenomena, such as the dynamics of small gas bubbles dispersed in the liquid phase, which instead sensibly reduce the quality of the time series of both optical and differential pressure sensors.

In addition to the intrinsic complexity of two-phase flows, it is difficult to compare results of reported experiments due to limited information on measurement techniques as well as on calibration and validation procedures. In particular, the QCVs calibration procedure [2, 7] does not appear appropriate for instantaneous local void fraction measurements (i.e., measurements detected in a portion of the pipe sufficiently short to be reliably approximated to a section). In fact, this technique offers an exact measurement of the void fraction by trapping a portion of the flow in a clear tube. The relative amounts of air and water in the tube can then be directly measured to obtain a void fraction value to be compared to the mean value of the void fraction time series measured from the sensor over a time interval corresponding to the valves distance. If the closing time of the valves is less than 0.01 s the experimental error can be neglected [7]. In addition, the portion of the pipe between the two valves must be long enough to neglect the internal volume of the two valves. On the other hand, the QCVs approach does not allow the estimation of a local value of the void fraction; in other words, this value cannot be directly correlated with the time series measured by the sensor, but only with its mean value evaluated in a time window specified as the ratio between the axial length of the calibration section (i.e., the distance between either the QCVs or the pressure tips) and the flow pattern characteristic velocity.

The aim of the present study is to present a probe for void fraction measurement and the procedure used for its calibration. In particular, an impedance sensor of the resistive type has been specifically designed and realized with the aim of achieving high-spatial and temporal resolution. Such resolution is considered fundamental in view of the future adoption of non-linear time series analysis techniques for advanced characterization of the dynamics of two-phase flows. The proposed calibration procedure is based on the estimation of the void fraction directly from high resolution photos. The following section is devoted to the description of the experimental setup and is followed by the description of the characteristics of the sensor, the calibration procedure and the experimental campaign, reproducing the main flow patterns families and allowing to test the performance of the system.

EXPERIMENTAL APPARATUS

Figure 1 represents a scheme of the experimental apparatus set up for the present study.

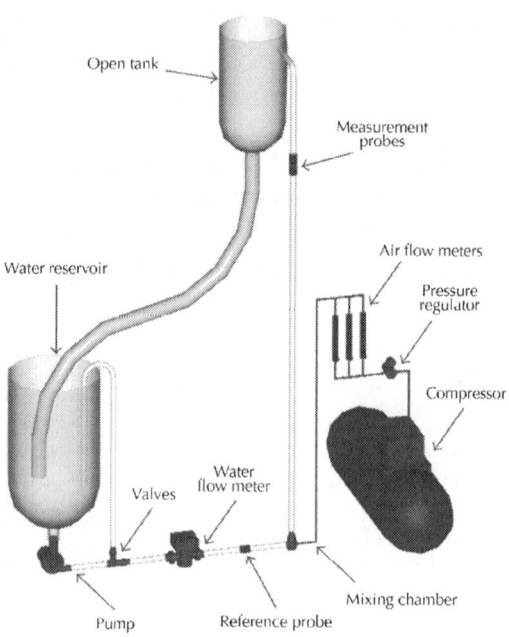

Figure 1: Experimental apparatus.

The liquid is supplied by means of a pump connected to a reservoir. The liquid flow rate can be varied up to 150 L/min by means of a series of valves and bypasses placed at the outlet of the pump. An electromagnetic flow meter is used in order to measure the velocity and the mass flow rate of the water. The liquid is distilled water with sodium chloride (1.5 g/L) in order to assure satisfactory conductivity. The temperature of the water in the reservoir has been maintained at 22°C (±0.2°C) in order to avoid the influence of temperature variation on conductivity measurement.

The airline is constituted from a pressure regulator (0–8 bar) and three air flow meters that can regulate the air flow rate in the range 0–200 L/min. The air is supplied to the mixing chamber by a pressurized tank fed by a compressor.

In order to allow the inspection of the flow pattern, the test section is constituted by a transparent vertical pipe of length 3 m and diameter 0.24 m. At its basis there is a mixing chamber that connects the liquid and air lines. In order to allow the degassing of the working fluid an open tank is placed on top of the vertical pipe.

Two void fraction probes are placed at a distance of 5 cm from each other; the lowest of them is at 2.40 m from the mixing section, that is, at a distance greater than the entrance region (assumed 80 times the pipe diameter [12]) in order to assure a well-established flow regime at the measurement section. The two probes permit to estimate the velocity of bubble during the calibration procedure. A reference probe is placed upstream the mixing chamber on the liquid line. All the probes are connected to an electronic circuit and an appropriate data acquisition system.

Design and Calibration of a Void Fraction Resistive Probe

If a significant difference exists in the electrical properties of the two phases, the volume fraction of one of the phases in a two-phase mixture can be determined by measuring the mixture impedance. The impedance is made up of both resistance and capacitance; therefore, a main point in the design of an impedance void fraction sensor is the choice of the excitation frequency, which determines the dominance of the resistive or capacitive behavior.

A mathematical model of an impedance probe can be briefly described as follows [13]. The impedance of a flow medium, Z, can be measured by two electrodes and can be expressed as:

$$Z(f) = \left[\frac{1}{R + \left(1/i2\pi f C_p \right)} + i2\pi f C_d \right],$$

(1)

Where R is the fluid resistance, C_p is the capacitance due to the polarization of the fluid molecules at the electrodes, C_d is the dielectric capacitance of the fluid, and f is the excitation frequency at the electrodes. In the case of a two phase mixture, R is a variable that depends on the difference of conductivity of the two phases (e.g., water and air for the present study). If the conductivity and the dielectric constant of the less-conductive phase (air) can be assumed negligible with respect to those of the conductive phase (water), C_p depends on the dielectric constant of the conductive phases whereas C_d is a function of the difference in the dielectric constants of the two phases, of the excitation frequency and of the void fraction. Therefore the void fraction can be determined by measuring either the resistance R or the capacitance C_d and, therefore, impedance sensors are usually classified as either conductive or capacitive sensors.

When the conductivity of the conductive phase is large, the measurement of capacitance C_d requires the use of high excitation frequencies (over 1 MHz), in order to minimize the role of the resistance and eliminate the parasitic capacitance caused by polarization at the electrodes, as well as to reduce the effect of external disturbances on the measurement system. Nonetheless, in order to overcome complications in the electronics associated to high excitation frequencies, the measure of the resistance is preferred. In this case the influence of C_d on measured impedance can be minimized by keeping the excitation frequency in a range depending on the electrical properties of the conductive phase. When water is considered as conductive phase such a range is 10–100 kHz. The choice of values higher than about 15 kHz is preferred in order to eliminate the role of parasitic capacitance C_p.

The sensor that has been designed and realized for the present study is schematized in Figure 2. The probe design is similar to that proposed in [2, 14, and 15]. It is made up by a pair of measuring electrodes that have been realized by means of gold-plated wires with a diameter of 0.6 mm forming two half rings facing one another, each spanning an arc of 90°. In addition, two pairs of gold-plated guard electrodes (identical to the previous) have been placed at a distance of 1 mm from the measuring pair on both its sides and maintained at the potential of the corresponding measuring electrodes. All the electrodes have been flush-mounted, in order to avoid disturbing effect on the flow.

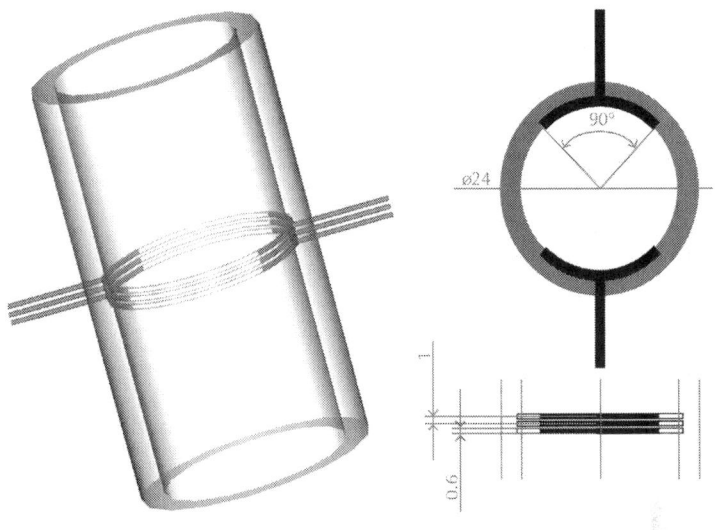

Figure 2: Schematic of adopted probes.

The criterion of geometrical design has been chosen in order to minimize the measurement volume. In fact, the axial length of the measuring electrodes can be neglected with respect to the pipe diameter so that the measurement volume can be approximated to the pipe transversal section. This means that the proposed electrode design allows the observation of local two-phase phenomena on the pipe section, ensuring higher spatial resolution than that obtained by other geometries of analogous sensors [16, 17]. It is worth observing that the sensor does not produce accurate measures of small-scale void fraction variations, in particular those characterized by a length scale below 0.6 mm. Such phenomena are typically associated to either disperse bubbles, occurring both in the bubbly flow and in regimes characterized by aeration of the liquid phase, or the high-frequency and low-amplitude components of the oscillations in wavy liquid films.

The probe has been operated in the resistive range. In fact, a carrier frequency of 20 kHz has been supplied by an external sine wave oscillator to both a measurement and a reference probe. The latter is a sensor identical to the measurement one that has been placed along the water line, as shown in Figure 1, in order to allow the elimination of the drift in void signals caused by possible changes in electrical properties of the flow medium.

Figure 3 shows the circuit used for signal processing. The instrumentation amplifier ensures high-dynamic response and perfect decoupling of the electronic circuit from the measuring section. The gain can be regulated by varying the R_G value. The amplified output is applied to the electronic rectifier. A cut-off frequency of 200 Hz has been adopted in order to remove the carrier frequency and to avoid aliasing with the sampling frequency. The final output is sent to a PC-based data acquisition system at the sampling rate of 1000 Hz, to allow the recording of the main void fraction fluctuations expected in the experiments.

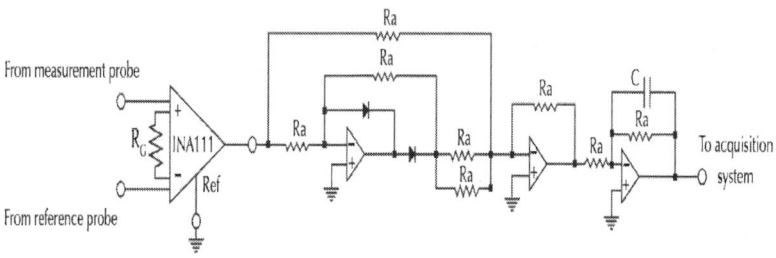

Figure 3: Electronic circuit for the probe.

In order to relate the resistive probe measurements directly to the actual instantaneous local void fraction, the calibration of the resistive sensor has been performed by comparing the value of the two-phase mixture conductivity measured by the sensor with the local diameter of the bubble as estimated by pictures taken by means of a high-resolution camera. In this way each point of the impedance time series describing the passage of a gas bubble has been correlated with the fraction of air occupying the pipe section. The estimation of the local bubble diameter requires the consideration of image distortion phenomena, deriving from the simultaneous presence of refraction and reflection at the curved pipe walls and of the liquid film around the bubble. Under the assumption of symmetric axial distribution of the interface between the two phases, which approximately holds for most of the flow patterns, distortion phenomena have been simultaneously estimated leading to the assessment of the correction function reported in Figure 4, which allows to evaluate the actual bubble local diameter.

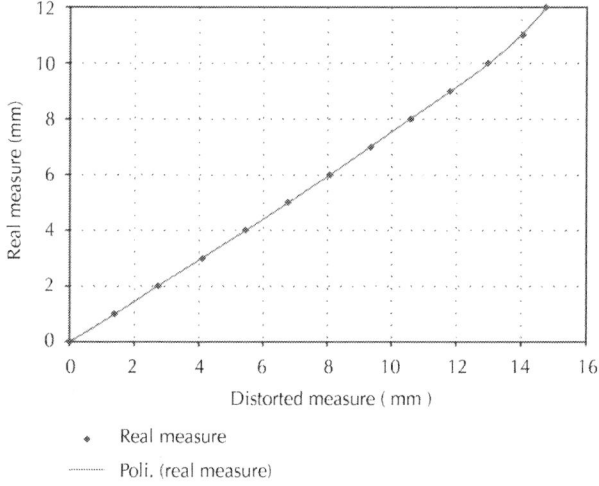

Figure 4: Correction function accounting for optical distortions.

After the determination of the correction function, it has been possible to calibrate the sensor response through the estimation of the instantaneous local bubble diameter from pictures taken by a high-resolution camera focused on the portion of the pipe immediately before the resistive sensor.

During the calibration procedure, the experimental apparatus has been operated setting up the air and water flow rates within the range of stability of the slug flow. After that the regime condition has been established in the system, the time series corresponding to the passage of a bubble through the measuring section of the resistive sensor has been recorded and compared with the synchronized picture of the bubble.

A code has been developed in order to process the pictures, performing the automatic extraction of the bubble contour and the correction of distortions. The code output is the spatial distribution of the void fraction along the bubble, which is related to the temporal distribution of the void fraction measured by the resistive probe by means of bubble velocity. The velocity value has been determined through the evaluation of the delay in the cross correlation function between the time series measured by two resistive probes placed at known distance (5 cm).

Figure 5 allows to compare the resampled time series of the resistive probe and the samples of the local void fraction as estimated from the measure of the bubble diameter. In this way each piece of the time series of the resistive probe, describing the passage of a bubble, has been correlated with the instantaneous value of the fraction of air flowing through the pipe section.

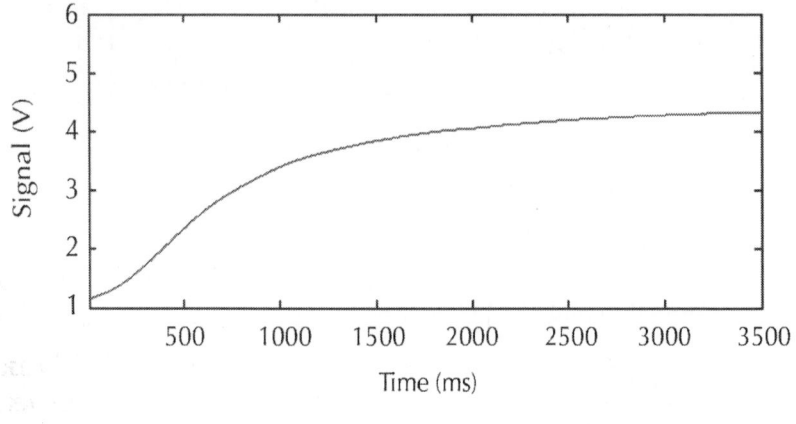

(a)

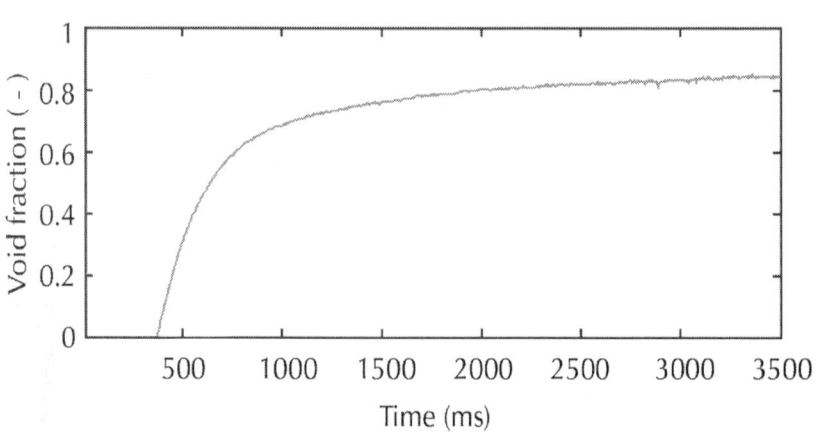

(b)

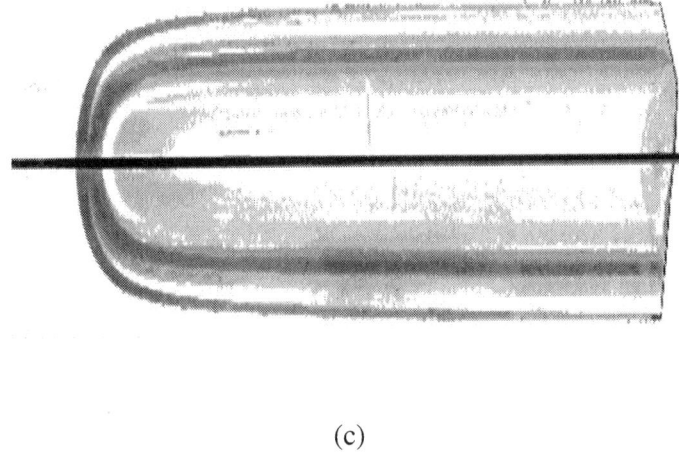

(c)

Figure 5: Comparison between the time series of the sensor (a) and the local void fraction estimation (b) obtained processing the picture of the bubble (c).

This procedure has been repeated for several samples and results have been summarized in Figure 6, by plotting the void fraction versus the conductance time series. Finally, the calibration curve reported in Figure 7 has been obtained as the seventh-order polynomial curve interpolating the estimated void fraction for several bubbles.

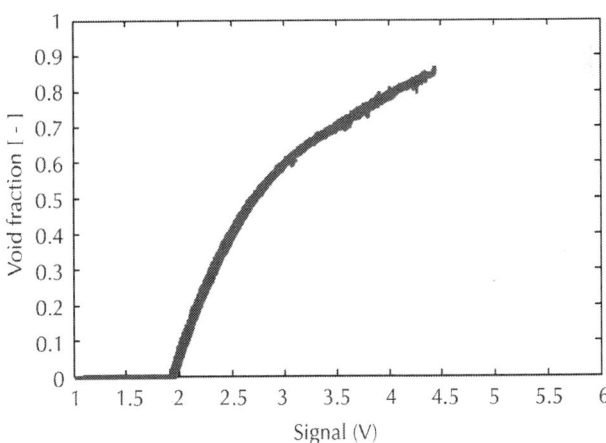

Figure 6: Calibration data for several bubbles detected under different operating conditions.

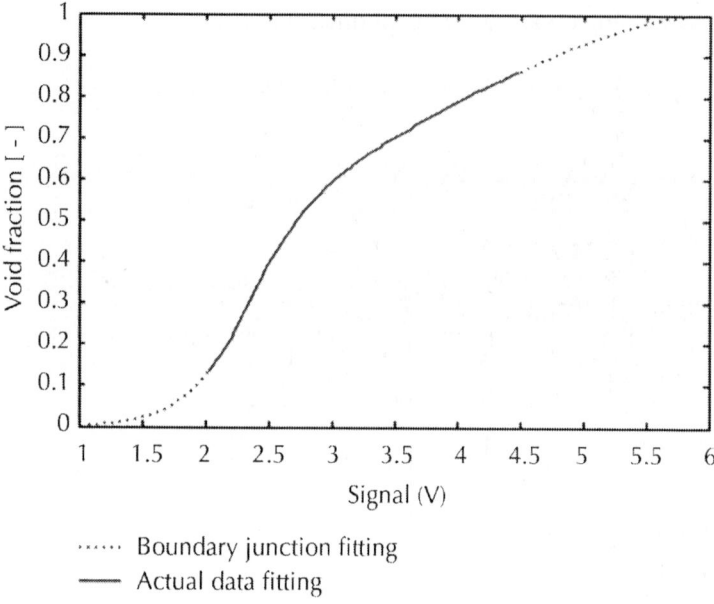

Figure 7: Calibration curve.

Experimental Tests

After the sensor characterization an experimental campaign has been performed. In particular, a series of tests has been carried out by varying the air and water mass flow rates, in order to identify the main types of flow patterns through the analysis of void fraction time series measured by the resistive probe. Table 1 shows the various operating conditions of the performed tests. Under these conditions different flow patterns have been observed to characterize the two-phase flow in the pipe.

Table 1: Operating conditions of experimental tests

						Air flow rate (lit/min)													
	1	2	3	4	5	6	7	8	9	10	15	20	30	40	50	60	70	80	90
	0.06	0.06	0.06	0.06	0.06	0.06	0.06	0.06	1.26	1.02	1.2	1.5	2.4	2.58	1.5	1.68	1.5	1.62	1.62
	0.9	1.2	1.8	1.2	1.02	0.72	0.9	1.86	2.88	3	3.84	3.84	3.84	3.78	1.92	2.28	2.1	2.4	2.16
	1.5	2.64	3.3	2.98	2.04	1.38	1.5	4.44	6.06	6.02	6.24	6.6	5.04	4.98	2.34	3	3.06	3.12	2.76
	2.1	4.38	4.68	3.72	3.18	2.58	3.12	5.7	9.3	6.06	7.02	9.6	6.18	6.54	3.3	4.14	4.26	4.08	3.9
	3.02	6	5.94	5.7	4.38	3.84	4.74	7.36	11.04	9.06	10.8	12.12	9.06	9.11	4.21	4.74	5.4	4.92	4.86
	3.61	7.35	7.3	7.08	6.06	4.86	5.82	9.9	15.0	11.88	12.6	15.42	11.4	10.98	6.42	6.48	6.18	5.58	6.3
	4.38	8.71	9.02	8.64	7.08	6.03	6.84	11.7	19.08	14.76	15	18.12	13.38	12.91	9.12	9.31	9.36	6.03	8.1
	6.01	10.35	10.74	10.08	9.76	7.26	8.11	14.11	21.84	18.2	18.1	21.84	16.26	15.36	10.98	11.04	12.18	9.03	9.72
	7.61	11.88	12	11.28	10.08	8.34	9.24	16.56	23.36	21.06	21.42	24.54	18.54	16.8	14.22	14.76	13.06	11.7	12.18
	9.06	13.74	13.44	12.3	11.88	9.48	10.44	18.42	28.98	24.6	24.6	27.9	22.08	18.6	17.52	17.76	18.24	14.4	15.06
	10.95	15.24	14.64	13.44	13.02	10.74	12	20.58	31.8	28.08	28.14	30.18	24.24	21.24	19.86	21	19.92	16.8	18.24
	12.24	16.2	14.92	14.7	14.1	12.18	13.62	22.26	33.06	39.1	29.98	33.9	27	23.58	22.08	23.04	21.6	19.68	21
	13.74	17.82	16.38	15.66	15.36	13.26	13.42	24.54	34.2	32.7	32.16	37.8	30.18	25.5	24.36	25.02	24.3	22.2	24
	14.64	19.8	17.82	15.9	16.08	14.7	16.2	26.4	35.88	34.56	35.76	40.02	33.18	27.75	27.24	28.38	27.06	24.6	27.18
	16.37	21.9	19.2	16.36	17.7	16.2	17.64	27.9	36.96	35.16	38.04	43.14	35.94	29.94	30.12	30.24	30	27.96	30.24
	17.04	21.3	20.64	18.12	18.6	17.28	18.38	32.58	37.62	36.84	39.42	46.74	38.52	33.12	32.82	33.36	32.88	30	33.06
	18.72	22.92	21.3	19.38	20.28	18.24	20.76	34.74	38.52	38.7	39.84	50.46	40.2	35.82	36.42	34.92	36.24	32.94	36.18
Water flow rate (lit/min)	19.08	24.6	23.22	20.58	21.78	19.26	22.14	35.82	41.1	41.82	43.02		40.98	38.28	39.24	37.08	39.18	36	39.18
	20.46	25.38	24.42	21.72	22.58	20.04	23.1	36.66	44.7	45.12	45.78		44.22	40.26	40.2	39.36	44.1	38.4	42.36
	22.38	26.58	25.08	23.58	23.88	22.26	24.3	37.92	47.16	48.86	48		46.68	43.86	43.26	45.36	47.52	40.44	45.36
	23.76	27.90	26.88	25.2	25.14	24.12	25.5	40.98	50.04	50.52	50.89		50.7	47.46	46.38	47.7		43.8	
	24.48	28.38	27.6	26.46	26.7	25.14	27.3	44.16							48.3	48.3		47.4	
	27.12	29.92	28.32	27.6	28.2	26.16	28.98	46.86											
	29.16	30.48	29.1	28.3	29.46	27.24	29.4	49.02											
	30.6	32.4	30.94	29.4	30.24	28.44	31.14												
	32.16	34.56	31.14	30.66	30.84	29.4	32.1												
	34.98	37.98	31.98	31.74	31.56	30.06	33.06												
	36.12	41.58	32.38	32.88	33.24	31.14	34.26												
	38.28	42.6	33.78	33.66	34.2	32.04	35.28												
	39.72	45.72	37.92	34.8	35.76	33.24	36.54												
	41.58	49.38	40.2	36.06	36.66	34.14	37.38												
	43.56		43.02		39.12														
	45.3		46.26		42.18														
	46.86		49.8		45.42														
	49.38				47.16														
					49.5														

In particular, for the sake of the result presentation, Table 2 reports the correspondence that has been established between typical flow patterns as reported in [2] and a restricted selection of testing conditions in Table 1.

Table 2: Selection of testing conditions corresponding to typical flow patterns

Flow pattern	m_{Water}(lit/min)	m_{Air}(lit/min)
Bubble	30.60	1
Cap	10.98	1
Slug	2.40	1
Plug	2.40	10
Churn	3.06	40
Annular	1.62	90

VOID FRACTION TIME SERIES ANALYSES

This section reports the results of the preliminary analyses of the experimental time series detected by the void fraction probe. The aim of these analyses is to verify the capability of the void fraction sensor to deal with the time evolution of the complex dynamical phenomena governing the distribution of the two phases under the various flow patterns. For synthesis, results will be discussed only for the experimental conditions reported in Table 2, chosen as representative of the main classes of flow patterns. Nonetheless, reported considerations have been found to be of general validity for the entire set of conditions detected during the experimental campaign.

At first, the experimental void fraction time series have been analyzed in the time domain, as reported in Figure 8. An important observation on the quality of these time series concerns the substantial lack of noise which, on the other hand, usually affects the experimental results reported in literature [2, 8, 15, and 18]. At the same time, the temporal evolution of the experimental time series appears sufficiently smooth, that is, the acquisition frequency of the void fraction sensor is high enough to guarantee an appropriate description of the phenomena. In consideration of the thin measurement volume achieved with the architecture of the void fraction sensor, it is therefore possible to claim the general satisfactory performances of the sensor with respect to its spatial and temporal response. As a matter of fact, low noise influence and good spatial and temporal response represent one of the main goals which the design and calibration of the void fraction sensor herein described wished to achieve.

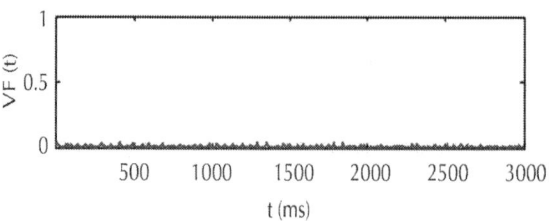

(a)

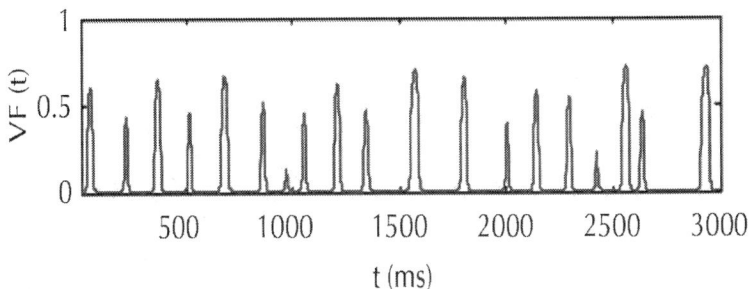

(b)

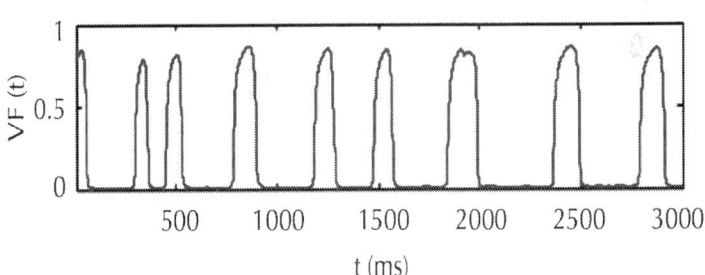

(c)

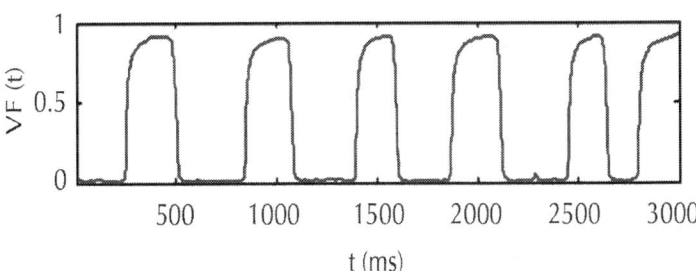

(d)

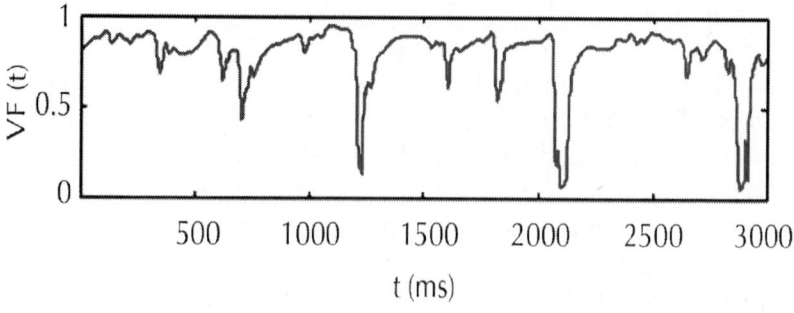

(e)

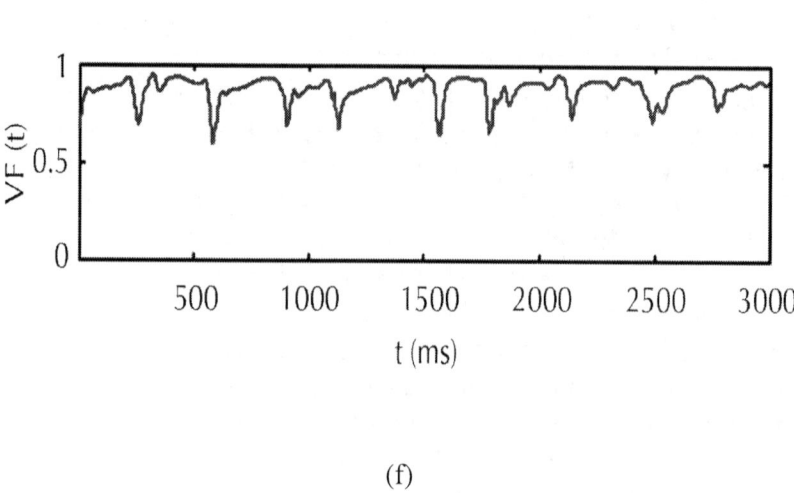

(f)

Figure 8: Experimental void fraction time series for the various flow patterns as reported in Table 2: (a) bubble flow; (b) cap flow; (c) slug flow; (d) plug flow; (e) churn flow; (f) annular flow.

Concerning the dynamical behavior of the time series reported in Figure 8, it is worth observing that, though several important differences exist between the time series of most flow patterns, with the exception of bubble flow, all of the time series are characterized by relevant amplitude and frequency differences between consecutive oscillations, showing the non-periodical nature of the system dynamics. Nonetheless, the repetition of similar waveforms seems to be a common feature of the various flow patterns, indicating that some kind of recurrent dynamics can be distinguished for each flow pattern. This result agrees with the analyses of the autocorrelation and of the power spectral density distributions of the experimental time series reported in a previous preliminary study [19] for analogous flow patterns detected with the same experimental apparatus. In [19] it was shown that, though a strong autocorrelation in the time series seems to indicate the deterministic nature of the experimental two-phase flow patterns, their complexity poses a limit to the validity of Fourier analysis and results in a typical broad-band power spectrum. As well known, both these observations hint at a deterministic source of chaotic dynamics, which cannot be satisfactorily described by means of previous linear tools.

Several studies [20–24] have claimed the existence of chaos in two phase flows but reported results are in general prone to relevant uncertainty due both to the general unsatisfactory spatial and temporal resolution of the experimental time series and to the strong influence of noise or of noise-like dynamics. Therefore, considering the satisfactory performance of the void fraction sensor described in this study in terms of both reduced noise influence and spatial and temporal time series resolution, [19] also reported the morphological analysis of the attractors of the experimental time series, evidencing the existence of a well-defined and regular structure in phase space, that is, a first important hint of deterministic chaotic behavior.

In consideration of the innovations introduced both in the sensor design and construction as well as in the calibration of the sensor, part of the experimental analysis aims at validating the performance of the sensor through the comparison with similar sensors. In particular, in the present study, the focus has been posed on verifying the correspondence between the statistical distributions of the measurements detected by the sensor with those reported in literature for the same kinds of flow patterns. In fact, the analysis of the statistical distribution of the experimental time series is indeed among the preferred tool to classify and distinguish the various flow patterns.

Figure 8 reports the distributions of the Probability Density Function (PDF) for the time series of the void fraction measurements detected during the experimental test reported in Table 2 and plotted in Figure 8. From the analysis of Figure 9 emerges that the application of PDF analysis to the time series detected during the experimental campaign allows a clear classification of two-phase flow patterns. In fact, for each flow pattern, the PDF is characterized by a well-defined distribution. Moreover, what is particularly relevant for the aims of the present study is that the PDFs calculated from the experimental time series well-correspond to the distributions reported in literature for the same kind of flow patterns [2, 3, and 15]. In other terms, the sensor performances are satisfactory not only in terms of spatial and temporal resolution and of low influence of noise, but also in terms of information content in view of flow pattern classification.

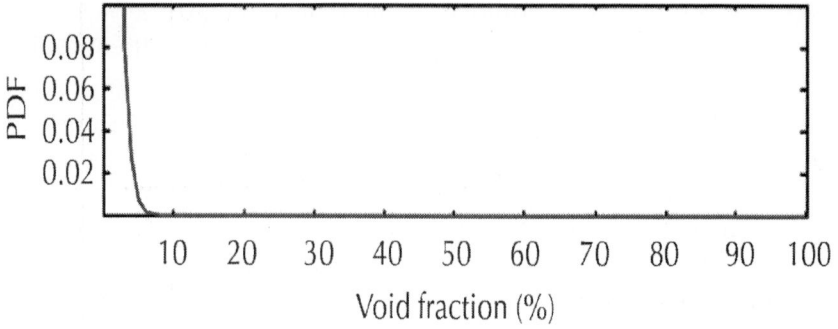

(a)

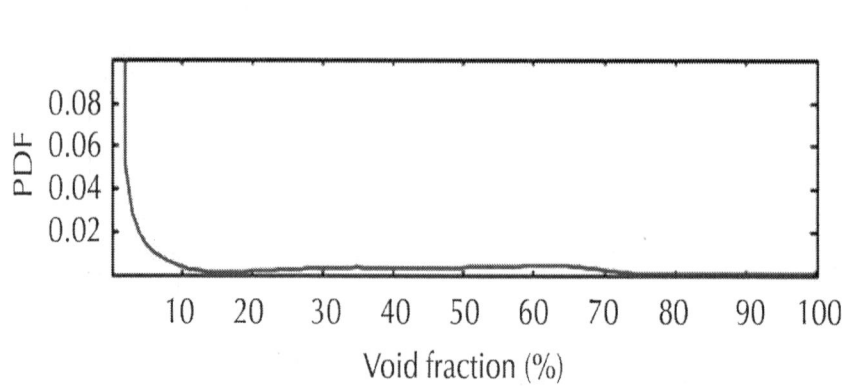

(b)

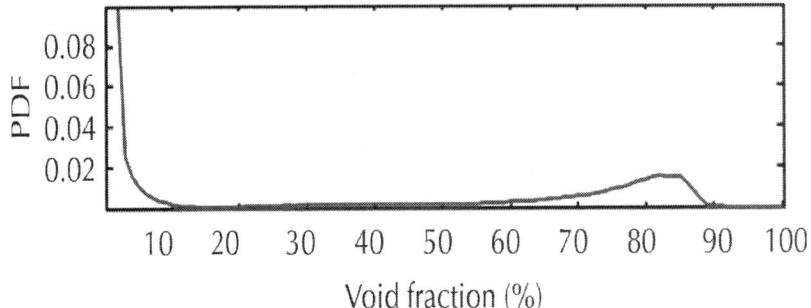

(c)

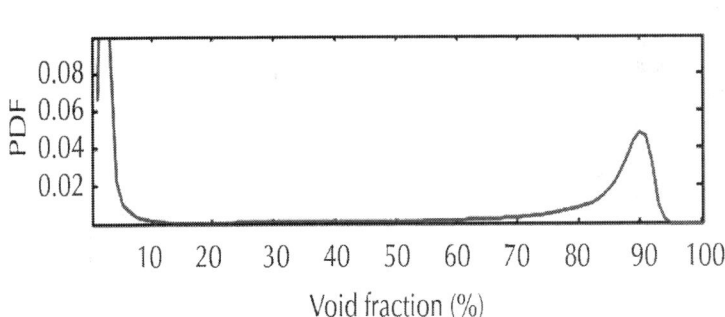

(d)

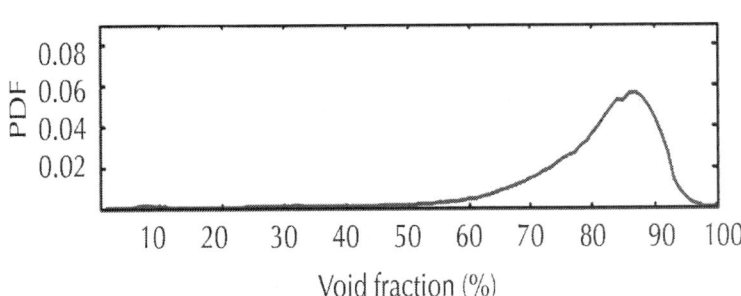

(e)

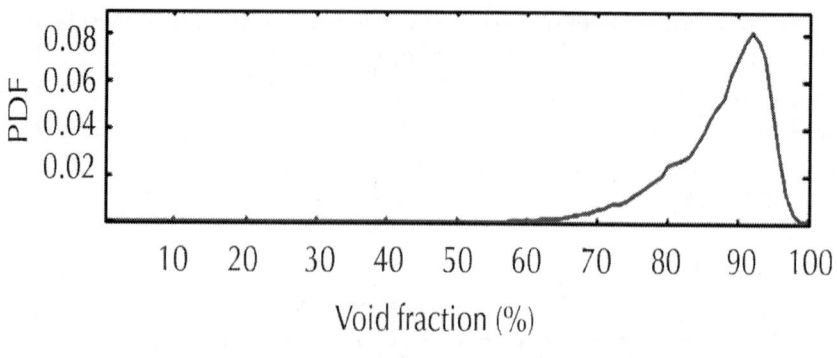

(f)

Figure 9: Distribution of the probability density function of the experimental void fraction time series for the various flow patterns as reported in Table 2: (a) bubble flow; (b) cap flow; (c) slug flow; (d) plug flow; (e) churn flow; (f) annular flow.

Therefore, future studies will be concerned on the exploitation of nonlinear tools both for the analysis of the experimental time series detected by means of the void fraction probe described in the present study and for the assessment of reliable criteria for flow pattern classification, based on an appropriate consideration of the complexity of the dynamics of two-phase flows.

CONCLUSIONS

This study is concerned with the design, construction, and calibration of an impedance probe for the measure of void fraction in air-water two-phase flows. The main goal in the definition of the sensor has been devoted to achieve high-spatial and -temporal resolution of the experimental void fraction time series, by means of appropriate geometrical design of the probe electrodes and electronic assessment of the data acquisition system. The adopted calibration procedure has been based on an estimation of void fraction through the implementation of a code for automatic extraction of bubble contours and the correction of image distortions.

For the entire spectrum of the possible flow patterns observed during the experimental campaign, the preliminary analysis in the time domain of the experimental time series detected by means of the void fraction probe, as well as the flow pattern classification based on the traditional PDF analysis, has pointed out the low influence of noise and the high-spatial and -temporal resolution of the experimental time series.

These are fundamental characteristics of the time series in view of the application of innovative tools of analysis oriented towards an accurate description of the complexity of nonlinear dynamical behavior characterizing two-phase flow.

REFERENCES

1. T. Taitel, D. Bornea, and A. E. Dukler, "Modelling flow pattern transitions for steady upward gas-liquid flow in vertical tubes," AIChE Journal, vol. 26, no. 3, pp. 345–354, 1980.

2. G. Costigan and P. B. Whalley, "Slug flow regime identification from dynamic void fraction measurements in vertical air-water flows," International Journal of Multiphase Flow, vol. 23, no. 2, pp. 263–282, 1997.

3. Y. Mi, M. Ishii, and L. H. Tsoukalas, "Vertical two-phase flow identification using advanced instrumentation and neural networks," Nuclear Engineering and Design, vol. 184, no. 2-3, pp. 409–420, 1998.

4. Y. W. Wang, B. S. Pei, and W. K. Lin, "Verification of using a single void fraction sensor to identify two-phase flow patterns," Nuclear Technology, vol. 95, no. 1, pp. 87–94, 1991.

5. R. Van Hout, L. Shemer, and D. Barnea, "Spatial distribution of void fraction within a liquid slug and some other related slug parameters," International Journal of Multiphase Flow, vol. 18, no. 6, pp. 831–845, 1992.

6. F. Devia and M. Fossa, "Design and optimisation of impedance probes for void fraction measurements," Flow Measurement and Instrumentation, vol. 14, no. 4-5, pp. 139–149, 2003. · ·

7. D. Lowe and K. S. Rezkallah, "A capacitance sensor for the characterization of microgravity two-phase liquid-gas flows,"

Measurement Science and Technology, vol. 10, no. 10, pp. 965–975, 1999. · ·

8.   J. K. Keska and B. E. Williams, "Experimental comparison of flow pattern detection techniques for air-water mixture flow," Experimental Thermal and Fluid Science, vol. 19, no. 1, pp. 1–12, 1999. · ·

9.   C. Vial, E. Camarasa, S. Poncin, G. Wild, N. Midoux, and J. Bouillard, "Study of hydrodynamic behaviour in bubble columns and external loop airlift reactors through analysis of pressure fluctuations," Chemical Engineering Science, vol. 55, no. 15, pp. 2957–2973, 2000. · ·

10.   G. P. Lucas and I. C. Walton, "Flow rate measurement by kinematic wave detection in vertically upward, bubbly two-phase flows," Flow Measurement and Instrumentation, vol. 8, no. 3-4, pp. 133–143, 1997. · ·

11.   N. D. Jin, Z. Xin, J. Wang, Z. Y. Wang, X. H. Jia, and W. P. Chen, "Design and geometry optimization of a conductivity probe with a vertical multiple electrode array for measuring volume fraction and axial velocity of two-phase flow," Measurement Science and Technology, vol. 19, no. 4, 2008. ·

12.   T. J. Liu, "Bubble size and entrance length effects on void development in a vertical channel," International Journal of Multiphase Flow, vol. 19, no. 1, pp. 99–113, 1993.

13.   C. H. Song, M. K. Chung, and H. C. No, "Measurements of void fraction by an improved multi-channel conductance void meter," Nuclear Engineering and Design, vol. 184, no. 2-3, pp. 269–285, 1998.

14.   Y. Ma, N. Chung, B. Pei, and W. Lin, "Two simplified methods to determine void fractions for two-phase flow," Nuclear Technology, vol. 94, no. 1, pp. 124–133, 1991.

15.   M. J. Watson and G. F. Hewitt, "Pressure effects on the slug to churn transition,"International Journal of Multiphase Flow, vol. 25, no. 6-7, pp. 1225–1241, 1999. ·

16.   P. Andreussi, A. Di Donfrancesco, and M. Messia, "An impedance method for the measurement of liquid hold-up in two-phase flow," International Journal of Multiphase Flow, vol. 14, no. 6, pp. 777–785, 1988.

17. N. A. Tsochatzidis, T. D. Karapantsios, M. V. Kostoglou, and A. J. Karabelas, "A conductance probe for measuring liquid fraction in pipes and packed beds,"International Journal of Multiphase Flow, vol. 18, no. 5, pp. 653–667, 1992.

18. H. Yeung and A. Ibrahim, "Multiphase flows sensor response database," Flow Measurement and Instrumentation, vol. 14, no. 4-5, pp. 219–223, 2003. ·

19. L. Cantelli, A. Fichera, I. D. Guglielmino, and A. Pagano, "Nonlinear dynamics of air-water mixtures in vertical pipes: experimental trends," International Journal of Bifurcation and Chaos, vol. 16, no. 9, pp. 2749–2760, 2006. ·

20. J. Drahos, J. Tihon, C. Serio, and A. Lübert, "Deterministic chaos analysis of pressure fluctuations in a horizontal pipe at intermittent flow regime," The Chemical Engineering Journal, vol. 64, no. 1, pp. 149–156, 1996.

21. N. D. Jin, X. B. Nie, Y. Y. Ren, and X. B. Liu, "Characterization of oil/water two-phase flow patterns based on nonlinear time series analysis," Flow Measurement and Instrumentation, vol. 14, no. 4-5, pp. 169–175, 2003. ·

22. H. Letzel, J. Schouten, R. Krishna, and C. M. Van den Bleek, "Characterization of regimes and regime transitions in bubble columns by chaos analysis of pressure signals,"Chemical Engineering Science, vol. 52, no. 24, pp. 4447–4459, 1997.

23. S. F. Wang, R. Mosdorf, and M. Shoji, "Nonlinear analysis on fluctuation feature of two-phase flow through a T-junction," International Journal of Heat and Mass Transfer, vol. 46, no. 9, pp. 1519–1528, 2003. ·

24. F. Franca, M. Acikgoz, R. T. Lahey Jr., and A. Clausse, "The use of fractal techniques for flow regime identification," International Journal of Multiphase Flow, vol. 17, no. 4, pp. 545–552, 1991.

Chapter 5

A Visual Tool for Accessibility Study of Pipeline Maintenance during Design

Chu-Hsuan Lee[1], Meng-Han Tsai[2], and
Shih-Chung Kang[3]

[1]Department of Civil Engineering, National Taiwan University, Taipei, Taiwan

[2]Center for Weather Climate and Disaster Research, National Taiwan University, Taipei, Taiwan

[3]Department of Civil Engineering, National Taiwan University, Taipei, Taiwan

ABSTRACT

Background

Pipeline maintenance is becoming an important issue in modern construction and building information model (BIM) research. An understanding of pipeline accessibility considerations in terms of operation and maintenance is essential for planning and management. Previous studies have highlighted the complexity of multi-pipes including mechanical, electrical and plumbing (MEP) pipelines and the importance of information visualization, but few have proposed a way to consider accessibility problems during operation and maintenance.

Methods

Therefore, this study develops a systematic method to evaluate accessibility with respect to pipeline maintenance. We first divided pipeline accessibility into three categories: (1) visual accessibility¿ the visibility for an inspector to view; (2) approachable accessibility¿ the difficulty for an inspector to approach; and (3) operational accessibility¿ the pipeline that can be operated by the inspectors. We created mathematical models and discussed the ergonomic details about each category. We then developed a user interface, VAO Checker, in which V, A and O stand for visual, approachable and operational respectively, to display visual information about pipeline accessibility. Through instantaneous analysis, the system visualizes the accessibility of the pipelines. We visually represent the intersection and union of these three categories to illustrate the varying accessibility of pipe elements.

Results

A usability test was conducted to validate the system ¿s effectiveness. The results of the usability analysis show that users have higher correctness when using VAO Checker than 2D plan drawing and 3D model, and they evaluate the performance of this tool better than 2D plan drawing.

Conclusion

Pipeline designers can benefit by using this tool to sketch a suitable traffic flow for engineers to investigate. Furthermore, the substantial amount of information saved in the layout database could be referenced for future optimization.

BACKGROUND

Pipeline design has become increasingly important in modern construction. Operation and maintenance requires consideration of accessibility in the design of the layout of plant pipelines. Previous research has noted that piping accounts for 20% of costs for the industry as a whole (Calixto et al. [2009]) and over 50% of the total detail-design labor hours (Park and Storch [2002]). All other activities of following detail design depend on piping and massive savings are achievable by utilizing good layout design and engineering practices.

Mechanical, electrical, and plumbing (MEP) pipes used to be supplemental facilities in construction. However, they have become necessary facilities, especially in nonresidential construction, such as hospitals, fire stations, and plants. Coordinating a MEP system is a tremendous challenge in engineering fields such as advanced technology, health care, and biochemistry industries (Khanzode et al. [2008]). Knowing how to arrange MEP systems appropriately is one of the most crucial aspects of the design phase (Riley et al. [2005]).

Maintenance is a crucial phase in these types of construction. Based on a statistics on expenses in a typical water treatment plant in year-2008, maintenance takes 15% possession of the expenses (Biehl and Inman [2010]). Moreover, based on some European and U.S. case study, maintenance has remarkable impact on at least one of the environmental aspects (Junnila et al. [2006]). A poorly designed pipeline layout design wastes space and materials. Moreover, it can cause difficulty or even danger during manipulation and management.

Literature Review

The literature reviewed for this study included findings and recommendations related to piping that can be categorized into three main groups: a pipe-routing algorithm, the integration of multi-pipes, and the visualization of pipeline design.

Pipe-routing Algorithms

Pipe-routing design is a subset of assembly design that conceives collision-free routes for pipes. A survey by Qian *et al.* ([2008])

categorized it into four fields: industrial plant pipeline layout design, circuit layout design, aircraft design, and ship piping system design (Qian et al. [2008]). Several studies have been devoted to routing algorithms, and mainly focus on physical constraints that connect the terminals of given locations and avoid all obstacles. They then use economic constraints to minimize the length of pipes and the number of pipe turns, which leads to an optimal specification. However, few, if any, solutions have considered pipeline accessibility in relation to operation and maintenance. Table 1 shows previous studies have disregarded some important constraints (Guirardello and Swaney [2005]; Ito [1999]; Mitsuta et al. [1987]; Newell [1972]; Park and Storch [2002]; Rourke [1975]; Schmidt-Traub et al. [1998]; Wangdahl et al. [1974]; Zhu and Latombe [1991]). Zhou and Yin ([2010]) emphasized that practical constraints, such as maintenance requirements and manufacturability, are not well recognized, and how humans still play an important role in guiding the computer to finish the design (Zhou and Yin [2010]).

Table 1: Earlier studies of pipe routing

Algorithm	Network optimization				Maze		Escape	Genetic
Author	Newell	Wangdahl	Zhu	Guiradello	Rourke	Mitsuta	Schmidt	Ito
Year	1972	1974	1991	2005	1975	1987	1998	1999
Dimension	3D	2D	2D/3D	3D	3D	3D	3D	2D
Domain	General	Ship	Robotics	Plant	General	General	Plant	General
Operation/ maintenance	-	-	-	-	-	-	-	-
Installation	-	-	?	-	-	-	?	?
Safety	-	-	-	-	-	?	-	-

-: not considered, ?: partially implemented, ?: fully implemented.

Lee et al.

Lee et al. Visualization in Engineering 2014 2:6, doi: 10.1186/s40327-014-0006-y

Integration of Multi-pipes

An industrial plant typically has more than one kind of pipeline. Feng *et al.* ([2012]) indicated a large number of pipelines, multifarious design constraints, and numerous obstacles in layout complicate the design of a pipeline system (Feng et al. [2012]). Recently, engineers have mainly used existing CAD software for design assistance, which has increased the problems associated with experts, such as complex operation, a long design cycle, and low efficiency. Some researchers advocated a new layout space model to reduce high complexity and design interference in the automated design of pipeline systems (Deliang and Huibiao [2009]; Feng et al. [2012]). Kim *et al.* ([1996]) found the range and complexity of the constraints limits the possibility of automatic pipe route design, and demonstrated a more natural and effective representation for route optimization (Kim et al. [1996]). Previous studies recognized the complexity in pipeline arrangement and proposed some methods to reduce it (Biehl and Inman [2010]; Guirardello and Swaney [2005]). However, in many instances the pipeline layout cannot be simplified, so the complexity should be taken into account.

Visualization Regarding Pipeline Accessibility

Some researchers have begun noticing the utility of information visualization for construction purposes as a means of improving the data-rich, but information-poor, problems of the construction industry (Songer et al. [2004]; Tsai et al. [2010]; Tsai et al. [2013]). Previous research focused on the visualization of construction data, noting how it can help identify potential causal relationships among construction data (Korde et al. [2005]; Kuo et al. [2011]; Russell et al. [2009]). Gao *et al.*([2006]) investigated colored construction drawing, which can increase the efficiency and accuracy of communication between designers and contractors (Grootjans [2009]a). Chang *et al.* ([2009]) and Chen *et al.* ([2013]) suggested a systematic procedure to determine the most suitable colors for effectively presenting the construction information (Chang et al. [2009]; Chen et al. [2013]). This procedure includes the selection, evaluation, and testing of colors to ensure they match the meaning of the construction information with the cognition of the users. Wang ([2011]) used the conception of visualization to develop an approach for assessing reachability of wheelchair users (Wang [2011]). With reference to pipeline arrangement, Deliang and Huibiao ([2009]) pointed out that visualization can help handle the detection and response to collisions between pipes and obstacles (Deliang and Huibiao [2009]).

Expert Interviews

During the early stage of this research, we interviewed six experts to determine the requirements of pipeline design. They are all in the field of plant pipeline design, including three engineers from a construction company, two managers from a microelectronics corporation, and one executive officer from the Building Information Modeling (BIM) research center. After combining the opinion of experts with previous literature review, we mainly focused our research on pipeline accessibility during operation and maintenance, which is rarely discussed in previous studies.

Needs Analysis

We determined from the interviews that there are four main considerations in pipeline design: (1) the manufacturing process, (2) operation and maintenance, (3) cost, and (4) aesthetics. In a typical plant engine room, as depicted in Figure 1, the engineers first have to deliberate how the pipelines go according to the manufacturing process, which will influence productivity and efficiency. They then contemplate how the workers will handle the equipment, meters, and valves during the operation and maintenance phase. Cost and aesthetics are aspects used to optimize the consequences of designs. Previous studies have proposed many algorithms by considering the cost factor, but maintenance is rarely discussed.

Figure 1: A typical pipeline arrangement in an engine room.

We mainly focused on operation and maintenance. Pipeline accessibility is the key factor to effective maintenance as it determines how easily the engineers can stretch to the accessories related to pipelines, including equipment, meters, and valves. Engineers can sometimes see pipelines from a distance, but cannot approach them due to the obstacles in the way of the pipelines. In other cases, engineers cannot read the meters in detail or operate the valves without difficulty, because these parts are mounted too high. We seek an easy way to illustrate pipeline accessibility with a view to engineers benefiting from this intuitive tool during the construction cycle (i.e., design, operation, and maintenance).

Objective and Scope

The aim of this study is to develop a method to assist decisions about pipeline maintenance. One major challenge of coordinating MEP multi-pipes is identifying the spatial conflicts between systems. Through instantaneous analysis, the system automatically produces visual information indicating how much pipe access the engineers can have. This tool allows users to view, explore, and interact with the pipeline information via a direct manipulation interface in order to identify the spatial accessibility in a more intuitive manner. The user can thus obtain a comprehensive understanding of pipeline maintenance.

METHODS

We use a Venn diagram[a], a diagram that shows all possible logical relations between different sets, to differentiate three categories of pipeline accessibility. We then apply each section of the diagram to different scenarios. We further develop mathematical models and discuss the ergonomic details about each different category.

[a]Lewis, Clarence Irving and Leibniz, Gottfried Wilhelm (1918). A survey of symbolic logic, University of California Press.

Overall Procedure of Pipeline Accessibility

We proposed three categories, *visual*, *approachable*, and *operational* to present the extent to which the pipe elements are accessible. As shown in Figure 2, we use the intersection and union of these three categories to discuss different scenarios as follows:

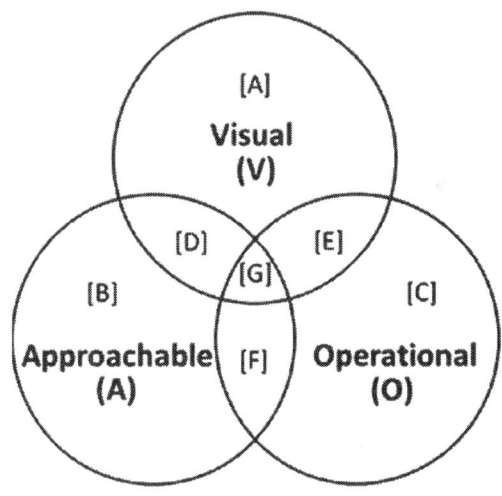

Figure 2: Venn diagram of pipeline accessibility.

Visual (V): determines how much of the pipe is directly visible for inspection.

Approachable (A): determines how far maintenance engineers can walk along the pipes.

Operational (O): checks how much of the pipes can be reached in order to operate valves or check surfaces.

In the Venn diagram, there are seven sections among the three circles. Each one is a variation of intersection and union. As listed in Table 2, we give the accessibility description of each variation from Figure 2.

Table 2: Seven variations of intersection and union

Section	Math representation	Accessibility description
[A]	V???A???O	Only visible, but not approachable and operable. This happens when obstacles and other pipes prevent engineers from accessing equipment and pipelines.
[B]	A???V???O	Only approachable, but not visible and operable. This happens when obstacles and other pipes block displays and controls.
[C]	O???V???A	Only operable, but not visible and approachable. Although remote control is possible, we did not consider this variation.
[D]	V ? A???O	Visible and approachable, but not operable. This happens when controls or valves are mounted too high, too low, or too far away to reach and operate.
[E]	V ? O???A	Operable and visible, but not approachable. The same as [C]. We did not consider this variation.
[F]	V ? O???V	Approachable and operable, but not visible. This happens when controls and valves are mounted behind the display, and engineers have to bend their arms to operate them. However, any blindness operation is not allowed in our assumption.
[G]	V ? A ? O	Visible, approachable, and operable¿the ideal situation.

Lee et al.

Lee et al. Visualization in Engineering 2014 2:6, doi: 10.1186/s40327-014-0006-y

These three categories are expressed in a visual conception of information. We adopted the anthropometric data from the American Bureau of Shipping (The American Bureau of Shipping[2003]) to build the model for accessibility analysis. We made some modifications by considering the physical differences between Americans and Taiwanese, because the first case would be a semiconductor fabrication plant in Taiwan.

Approachable Accessibility

This level determines how far people can walk along the pipes. Walkways should have 2.1 m minimum clearance above the walking surface for the full length and width of the walkway. The analysis and mathematical model of approachable accessibility is different from the other two because it is a dynamic process. As shown in Figure 3 and Table 3, we first use a bounding cylinder to represent a person, and bounding boxes in different sizes to represent a cart in different applications. If obstacles or other pipes block the box, it cannot go farther along the pipes.

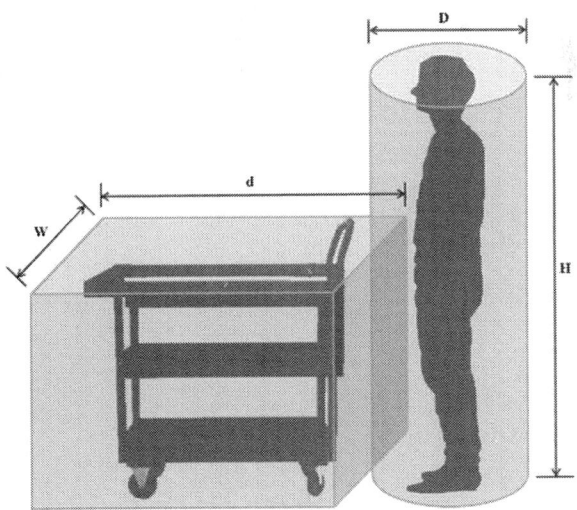

Figure 3: Bounding cylinder and box representation.

Table 3: Bounding box size for recommended walkway dimensions

Application	Box size*
One person traveling in an area with limited access	51×51×150
One person in unrestricted area, where two persons could pass	71×71×210
One person with a cart	71×120×210
Normal two-way traffic or any means of egress that leads to an entrance or exit	92×120×210
Corridor or passageway that serves as a required exit	112×120×210

*Size representation: W (cm) × (D + d) (cm) × H (cm).

Lee et al.

Lee et al. Visualization in Engineering 2014 2:6, doi: 10.1186/s40327-014-0006-y

The mathematical model of visual accessibility is then constructed as the equation:

$$A = (H, r, P) \tag{1}$$

As denoted in Figure 4, $r = \dfrac{Max(W,D)}{2}$, and we used a cylinder with radius r and height H to simplify the bounding box. S means the start point, and T means the target point. P is the path from S to T:

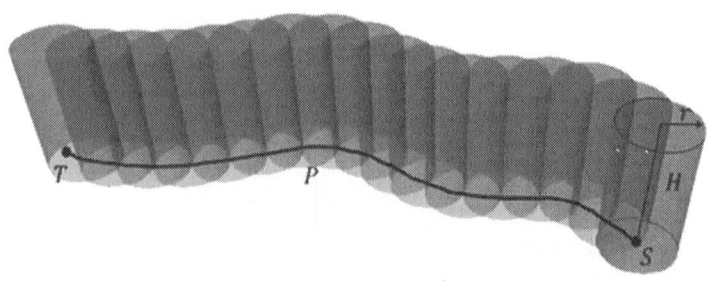

Figure 4: Mathematical model of approachable accessibility.

$P = [S, p_1, p_2, ..., p_n, P_{n+1}, ... T]$, where the cylinder is not blocked.

Visual Accessibility

This level determines how much of the pipe is directly visible for inspection. We further divide it into two levels: visible and legible. The former includes those used for normal operations and those not requiring accurate readings, whereas the latter includes those used frequently, for obtaining precise readings, and in emergencies. The mathematical model of visual accessibility is constructed as the following equation. Figure 5 indicates the parameters.

$$V = \left(S, {}^{o}H, {}^{o}L_{min}, {}^{o}L_{max}, \theta, {}^{o}H^{v}_{min}, H^{v}_{max}\right)$$

(2)

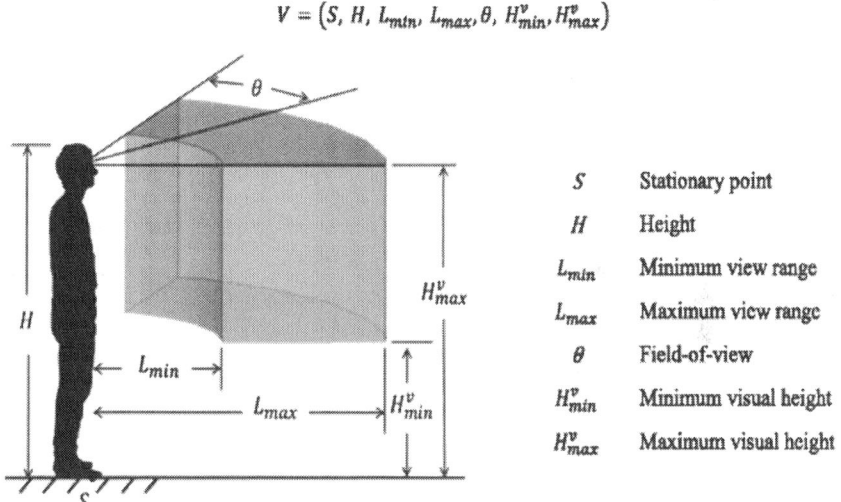

$$V = \left(S, H, L_{min}, L_{max}, \theta, H^{v}_{min}, H^{v}_{max}\right)$$

S	Stationary point
H	Height
L_{min}	Minimum view range
L_{max}	Maximum view range
θ	Field-of-view
H^{v}_{min}	Minimum visual height
H^{v}_{max}	Maximum visual height

Figure 5: Mathematical model of visual accessibility.

Figure 6 indicates people¿s field-of-view. The two parameters regarding it are the distance from eyes (L) and the viewing angle from the central line (?). Based on ABS research, as shown in Table 4, people can see the details of pipes at distances between 33 cm and 71 cm, and a viewing angle within 35 degrees, where the legible level should be located (provided obstacles or other pipes do not block the pipes and displays). The distance for the visible level can be up to 200 cm, with the viewing angle up to 60 degrees. Figure 7 illustrates the visual heights (Hv) for displays in different postures: standing (C), kneeling (D), and squatting (E). Table 5 shows the maximum and minimum heights for the legible and visible levels, based on personal height (H). Because the range of these three postures overlapped, we integrated the data. The legible level should be located within the multiple 0.4261-0.9375, but the visible level can be broader, 0.2955-1.0114.

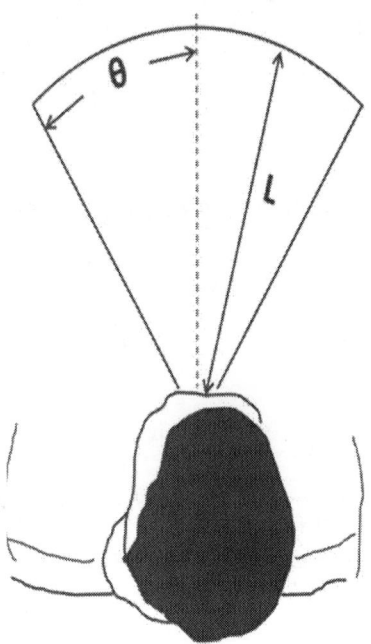

Figure 6: Field-of-view.

Table 4: Suitable range of field-of-view

	L (cm)	? (degrees)
Legible	33-71	35
Visible	0-200	60

Lee et al.

Lee et al. Visualization in Engineering 2014 2:6, doi: 10.1186/s40327-014-0006-y

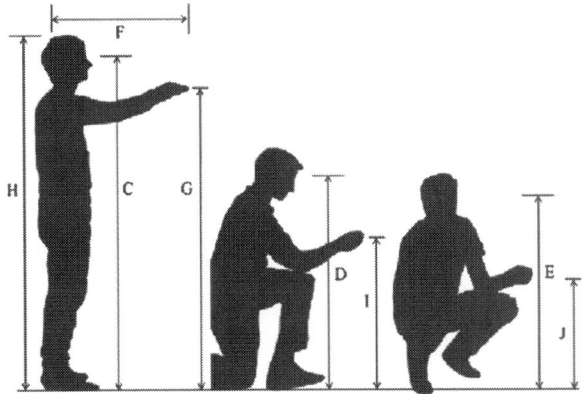

Figure 7: Related height in different postures.

Table 5: Suitable field-of-view and vision range (multiple of H) for legible and visible levels

Posture	L (cm)	? (degrees)	Standing (C)	Kneeling (D)	Squatting (E)	Overall
Visible maximum	200	60	1.0114	0.8239	0.7102	1.0114
Legible maximum	71	35	0.9375	0.7500	0.5795	0.9375
Legible minimum	33	0	0.7216	0.5398	0.4261	0.4261
Visible minimum	0	0	0.5909	0.3977	0.2955	0.2955

Lee et al.

Lee et al. Visualization in Engineering 2014 2:6, doi: 10.1186/s40327-014-0006-y

Operational Accessibility

To facilitate the operation of valves or the checking of surfaces, this level checks the accessibility of pipes. It is derived from the arrival accessibility level, and shows the ease with which people can operate within the pipe layout. We further divided it into two levels: general control and precise control. The former includes those used for normal operations and those not requiring accurate manipulation, whereas the latter includes those used frequently, for obtaining precise performance, or in emergencies. The mathematical model of operational accessibility is constructed as the following equation. Figure 8 indicates the parameters.

$$O = \left(S, {}^oH, {}^oF, {}^oH^o_{min}, H^o_{max}\right) \qquad (3)$$

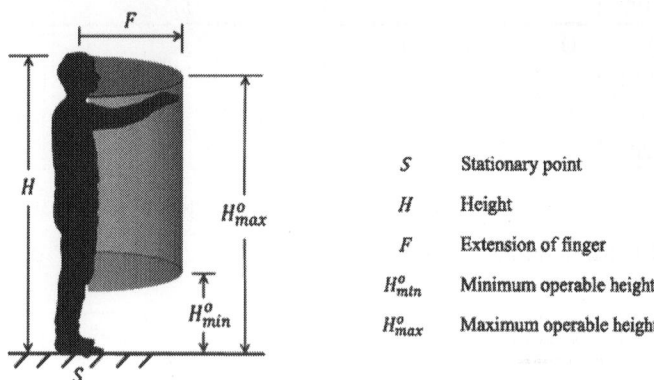

S	Stationary point
H	Height
F	Extension of finger
H^o_{min}	Minimum operable height
H^o_{max}	Maximum operable height

Figure 8: Mathematical model of operational accessibility.

People¿s forward functional reach from behind the shoulder to the tip of the extended finger (*F*) and the operable heights (*H0*) for controls in different postures are illustrated in Figure 7: standing (G), kneeling (I), and squatting (J). Table 6 shows the maximum and minimum forward functional reach and heights for precise and general controls, based on personal height (*H*). Frequently used controls should be located within a radius of multiple 0.2614 from the operator¿s centerline, whereas less frequently used controls should be located within a radius of multiple 0.4545 from the operator¿s centerline. Because the range of these three postures overlapped, we integrated the data. Precise control should be located within the multiple 0.2273-0.7670, but general control can be broader, 0.2045-1.0966.

Table 6: Suitable forward functional reach and heights (multiple of H) for precise and general controls

Posture	Forward (F)	Standing (G)	Kneeling (I)	Squatting (J)	Overall
General maximum	0.4545	1.0966	0.8239	0.7102	1.0966
Precise maximum	0.2614	0.7670	0.6136	0.4545	0.7670
Precise minimum	0	0.4886	0.3068	0.2273	0.2273
General minimum	0	0.4318	0.2614	0.2045	0.2045

Lee et al.

Lee et al. Visualization in Engineering 2014 2:6, doi: 10.1186/s40327-014-0006-y

Implementation

This study developed a system, VAO Checker, which integrated the user interface and visualization information as a tool, to implement the proposed methodology. The following sections describe the software used for the development environment and the system design.

Programming Platform

This study used Microsoft Windows Presentation Foundation (WPF) for the display of the user interface. WPF was chosen because it allows programmers to easily unify multimedia data, and change the appearance or the function of display controls for customization. Furthermore, the WPF application functions by off-loading to graphics processing units (GPUs) rather than central processing units (CPUs), which facilitates smoother graphics and better performance (Nathan [2006]).

Graphics Engine

The framework developed for the visualization information was based on the Microsoft XNA Game Studio 4.0. This tool assists the development of video games and the improvement of software management. XNA has ample performance for the development of 2D and 3D games. It offers users the capability to build the operating system and visual images with ease (Grootjans [2009]; Miller and Johnson [2010]).

System Design

The proposed tool called VAO Checker was built for this study to consider the three categories of pipeline accessibility. As shown in Figure 9, the operation interface displays a plan view of the space, including the equipment and pipelines. The user can use this tool to find a collision-free path through the space and to examine the different levels of visual and operational accessibility.

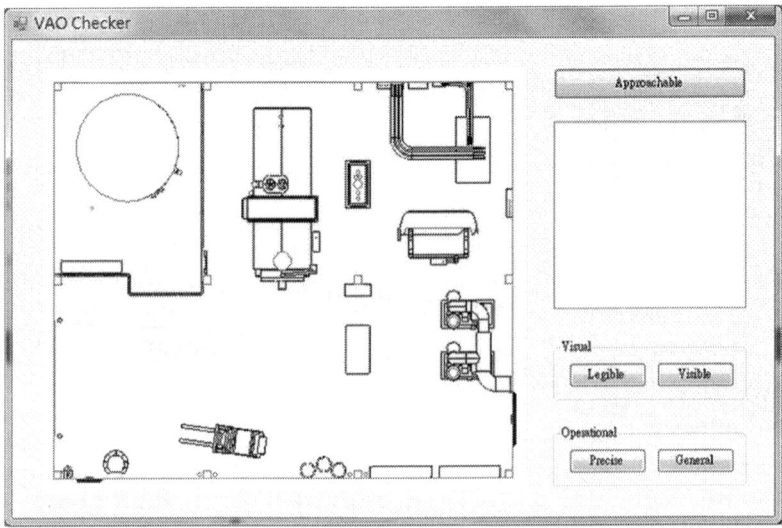

Figure 9: Operation interface of VAO Checker.

Example Case

We created an example case to validate the practicality of VAO Checker. We built a virtual building project as an example case in a machinery room (Figure 10), which has some basic equipment and a pipeline arrangement.

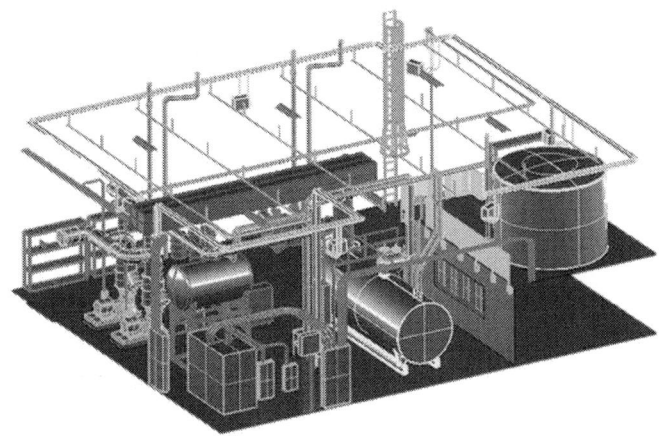

Figure 10: Example case.

After the start point and target point are decided, and the ¿Approachable¿ button is clicked, a collision-free path with some check points listed in the blank comes into view (Figure 11). The dot with sufficient approachable accessibility is bigger and colored light green, and the dot with limited approachable accessibility is smaller and colored dark green.

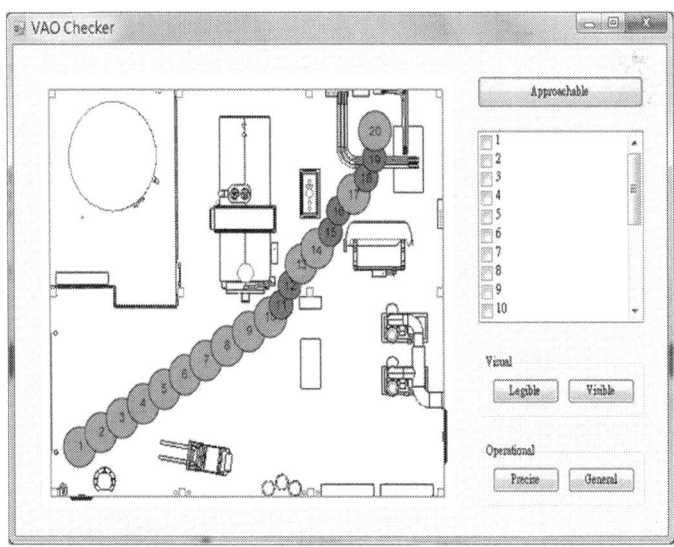

Figure 11: Collision-free path from start point to target point.

The user can choose one of those listed points, and the chosen point will turn into a red dot (Figure 12). The user can then examine different levels of visual and operational accessibility by clicking the four buttons at the bottom right corner. A visualization window, indicating a corresponding level of accessibility, will open (Figure 13). The user can utilize some specific keys to interact with the pipeline information, such as rotating the view direction or stepping forward or backward.

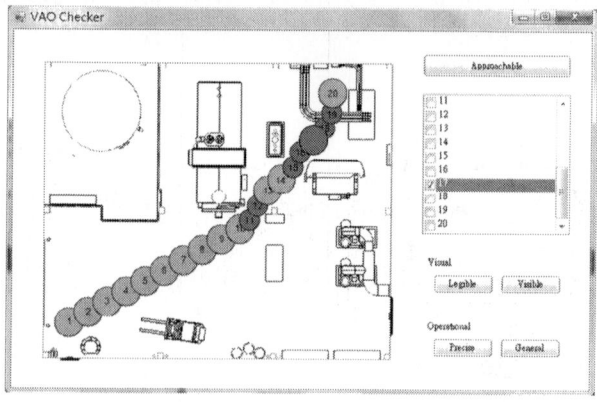

Figure 12: Chosen check point turns into a red dot.

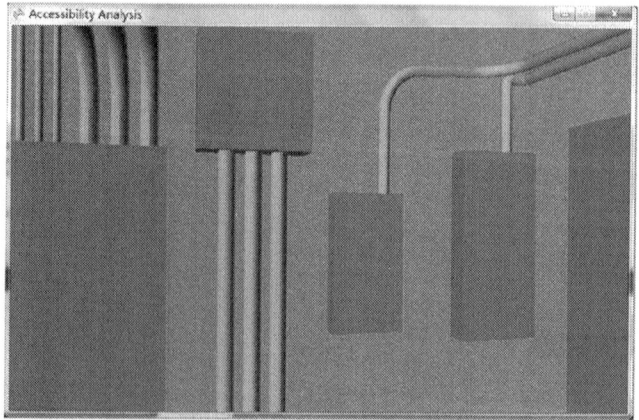

Figure 13: Visualization window indicating corresponding level of accessibility.

Validation

In order to verify how VAO Checker could help users explore and understand relevant accessibility information, we conducted a usability test. We also solicited expert consultation to verify the usability and how the users can interact with the pipeline accessibility information.

Test Plan

Test Procedure

For the usability test, we built a typical machinery room project with equipment and pipelines. There were 10 accessibility problems in this case, which are categorized in Table 7 according to Figure 2and Table 2. All users had to identify the problems in three individual tasks, each task using different mediums, 2D plan drawing, 3D model and our system, VAO Checker. Besides, we also conducted the NASA Task Load Index (NASA-TLX) test. As shown in Figure 14, the test plan began with the NASA-TLX weight assessment, in which the user compared the factors pairwise based on their perceived importance. After the user finished the identification of accessibility problems via one information medium, the user had to rate each factor of task load within a 100-points range. The final NASA-TLX score was calculated based on the weight distribution, which was decided at the initial phase.

Table 7: Category of 10 accessibility problems

Category	[A]	[B]	[C]	[D]	[E]	[F]	[G]	Total
Amount	3	2	-	3	-	2	-	10

Lee et al.

Lee et al. Visualization in Engineering 2014 2:6, doi: 10.1186/s40327-014-0006-y

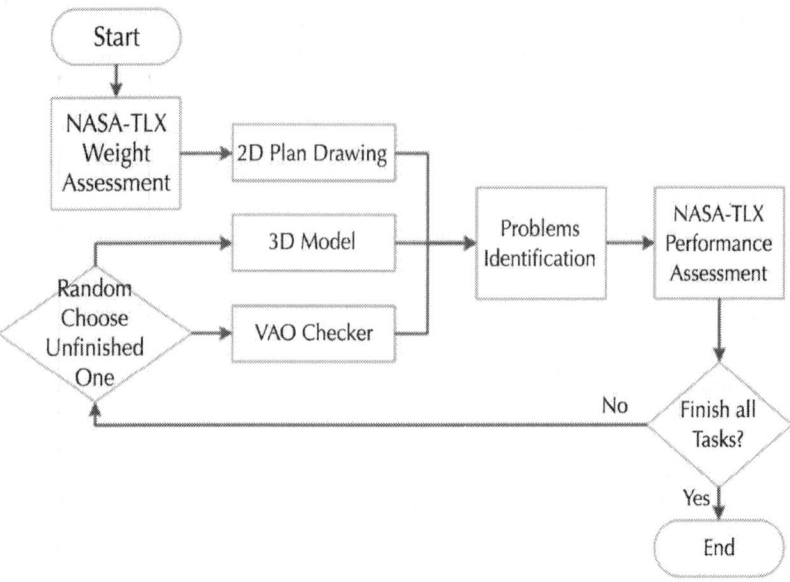

Figure 14: Usability test procedure.

Test Participants

There were 30 participants in the usability test, 19 male and 11 female. Their ages ranged from 20 to 37 years old. We solicited participant from non-engineering background as well, because they might provide suggestion from different point of view. Among the participants, 27 participants are from a civil engineering background, 2 from a psychology background, and 1 from an economics background.

Test Environment

The usability test was conducted in a controlled environment that was limited to the room shown in Figure 15. In this room, each participant was asked to sit at the east side of the front table in the room. A researcher, sitting next to the participant, conducted and facilitated the test procedure and guided the participants through the test.

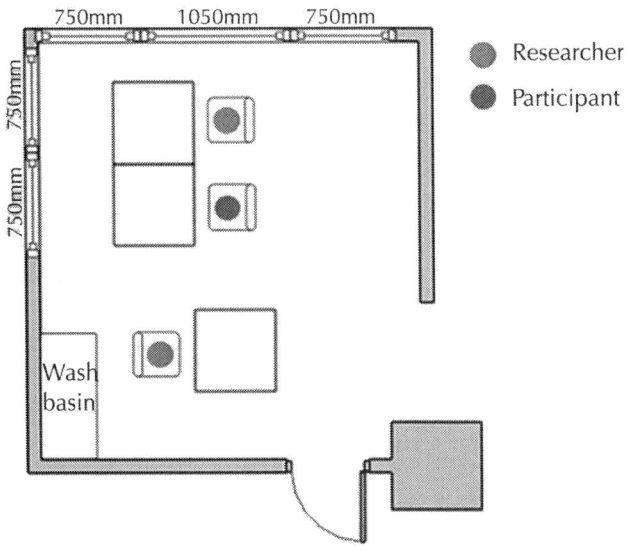

Figure 15: The test environment.

RESULTS AND DISCUSSION

An ? level of 0.05 was used for all statistical tests and analysis, and we calculated the p-value between groups in analysis of variance (ANOVA), where $p?<?0.05$ means statistically significant. The test results assessed how quickly and accurately participants performed the task when using different mediums. There is also an analysis of NASA-TLX score, which shows how the participants evaluated the ergonomics performance of each medium. They are summarized as follows:

Correctness: VAO? 3D > 2D

The box-and-whisker plot, a visual display of the five number summary, of success rate of each medium is shown in Figure 16. Table 8 presents means and standard deviations of success rate of each medium, and the p-value shows the data between 2D and VAO Checker is statistically significant. As the data indicates, the success rate of VAO Checker (64.3%) is 1.6 times higher than 2D plan drawing (40.1%) and 1.14 times higher than 3D model (56.4%).

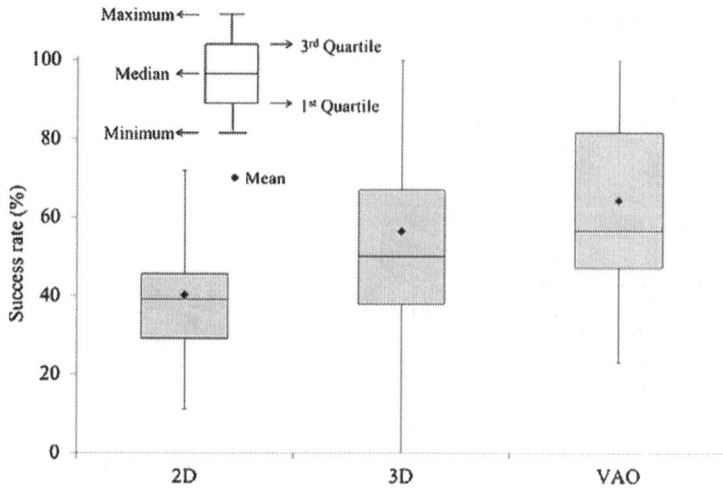

Figure 16: Success rate of each medium.

Table 8: Statistical analysis of correctness

Medium	Mean (%)	Std. Deviation (%)	p-value	
2D plan drawing	40.1	16.3	2D & 3D	0.002*
3D model	56.4	25.3	2D & VAO	0.000*
VAO Checker	64.3	24.5	3D & VAO	0.139

*the data is statistically significant.

Lee et al.

Lee et al. Visualization in Engineering 2014 2:6, doi: 10.1186/s40327-014-0006-y

Performance: 3D > VAO > 2D

The box-and-whisker plot of NASA-TLX score of each medium is shown in Figure 17. Table 9 presents means and standard deviations of NASA-TLX score of each medium, and the p-value shows the data between each pair of these three groups is statistically significant. The score of 2D plan drawing is the lowest (36.0), whereas the score of 3D model is the highest (53.8). The score of VAO Checker (48.0) is 1.33 times higher than 2D plan drawing.

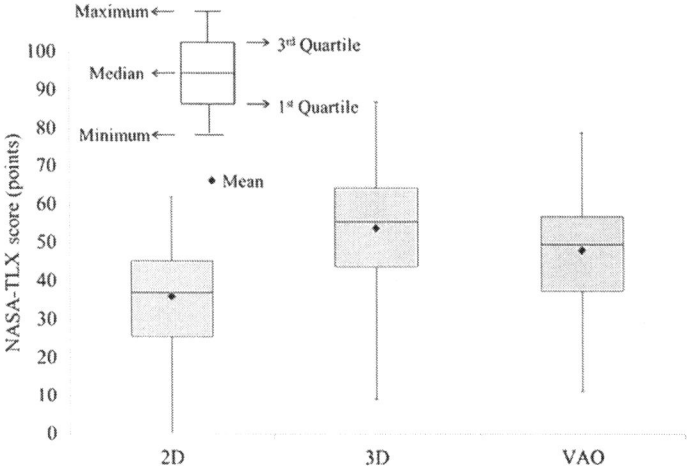

Figure 17: NASA-TLX score of each medium.

Table 9: Statistical analysis of performance

Medium	Mean (points)	Std. Deviation (points)	p-value	
2D plan drawing	36.0	13.5	2D & 3D	0.000*
3D model	53.8	17.0	2D & VAO	0.004*
VAO Checker	48.0	17.3	3D & VAO	0.020*

*the data is statistically significant.

Lee et al.

Lee et al. Visualization in Engineering 2014 2:6, doi: 10.1186/s40327-014-0006-y

Findings from the Result

Most of the participants have a background of civil engineering, and they can get on track quickly when they check 2D plan drawing or 3D model. Based on the observation during the usability test, participants would spend some time to get used to the user interface of VAO Checker, because it is a new tool for them. As a result, the average problem identification time of VAO Checker is longer than 2D plan drawing and 3D model.

However, in the analysis of correctness, the success rate of VAO Checker is the highest. This means, although users might spend more time when they first contact with the user interface of VAO Checker, they still can achieve the goal of high correctness. Some participants advised that in addition to the three categories of accessibility, VAO Checker should take more safety factors into consideration. They indicated that the section where steam is generated should be prohibited to pass through. Then, the path should bypass those areas.

In the analysis of performance, the NASA-TLX score of 3D model is slightly higher than VAO Checker. We also interviewed the participants about their feeling when they manipulated VAO Checker. Many of them pointed out that the manipulation of VAO Checker had a sense of reality, unlike 2D plan drawing. They could look around the environment, and perceive the size of equipment and pipelines. The visual effects made it like playing a game. However, because the viewing angle of VAO Checker is set as first person, they sometimes got confused with the direction in the virtual environment. On the contrary, the viewing angle of 3D model is set as third person, and they can identify the direction easily. That is the reason some participants evaluated the NASA-TLX score of 3D model higher.

Furthermore, many of the participants commented that another reason they got confused with the direction is the unfamiliarity with the overall pipeline design. VAO Checker would serve as a useful tool for the designers who are conscious of the design, and they would benefit from this tool to correct any design errors. They suggested that VAO Checker is suitable for planning a more complex environment, such as chiller machinery room. The sizes of pipelines are bigger, and there are more relevant systems. Formerly only experienced designers could plan a pipeline layout which is acceptable enough. Through VAO Checker, designers could save a lot of time in analyzing and planning.

Owing to the booming BIM industry, some participants supposed VAO Checker would be developed into an application-programming interface (API) of BIM related software. Construction companies or corporations are increasingly using BIM as a tool to integrate information from the fields of architecture, structural engineering, and MEP systems. If VAO Checker can be developed into an API, more pipeline designers and engineers can benefit by saving time and effort.

Despite the participants needed some time to be familiar with the manipulation interface of VAO Checker, they all agreed that they could identify the accessibility problems very easily via this tool, because it provided sufficient information for them to judge the level of pipeline accessibility. They expected the path generated from analysis of approachable accessibility could be used for inspection or judgment, and the engineers would have a certain understanding of pipeline maintenance of the entire environment if they could move along this path.

Contribution

Unlike previous studies, in which few solutions have been offered to propose a way to consider pipeline accessibility through maintenance, our research emphasizes the importance of pipeline accessibility and makes efforts on the following sections:

A Useful Tool for Pipeline Designers and Engineers

We have shown that VAO Checker serves as a useful system for pipeline designers and engineers during operation and maintenance. Designers can benefit by sketching a suitable traffic flow for the users¿ investigation. Engineers can obtain a comprehensive understanding of the pipeline maintenance with the aid of the interface and can identify spatial accessibility in a more intuitive manner.

Considering Pipeline Accessibility

Unlike previous studies, in which few, if any, solutions have been offered to propose a way to consider pipeline accessibility through operation and maintenance, our research emphasizes its importance.

Integration of Information from Multi-pipes

With regard to the complexity of multi-pipes, different kinds of pipes might have to be considered in terms of different levels of accessibility.

We developed mathematical models about each different accessibility category and discussed the ergonomic details.

Enhancing Comprehension via Visualization

Given the importance of visualization for pipeline accessibility, we developed VAO Checker, which integrated the user interface and visualization information as a tool to provide users with complete information about pipeline accessibility.

Future Work

Even though this research has made efforts on the pipeline accessibility, there are still some possibilities can be worked on in the future:

Number of Levels Divided for Each Accessibility Category

Operational accessibility, for example, has more than one kind of controls, such as toggle switches, pushbuttons, and rotary controls. Each demands a different level of sophistication, and might need a different description for the mathematical model of operational accessibility. From the standpoint of plan amendments, VAO Checker could be much more practical tool if the analysis result can show the segmented and numerical degree of accessibility.

Optimization via Operators

The system we propose uses computation to suggest a suitable path by considering approachable accessibility. It might become much more usable if the pipeline¿s designer can improve results through specific operators based on the designer¿s experience. Furthermore, the tremendous amount of information saved in the layout database might be referenced for future optimization.

Development of Pipe Assembly Planning

The pipe-routing design still relies on designer¿s experience. We hope this system can be extended to pipe assembly planning areas for efficient implementation, which might lead to a process of automatic pipe routing. The pipeline arrangement will only become more complex in the future, and pipe assembly planning will help increase the effectiveness and efficiency of routing design.

CONCLUSIONS

This research developed a systematic method to evaluate the accessibility of pipeline maintenance. During the early stage of this research, we interviewed six experts to determine the requirements of pipeline design. After combining the opinion of experts with a literature review, we mainly focused our research on pipeline accessibility during operation and maintenance, which is rarely discussed in previous studies. We first divided pipeline accessibility into three categories, developed mathematical models, and discussed the ergonomic details of each different category. We then developed a system called VAO Checker, which integrated the user interface and visualization information as a tool to implement the proposed methodology. VAO Checker used a simple motion-planning algorithm to find a path with acceptable approachable accessibility, and programmed the mathematical models into visualization information indicating the visual and operational accessibility. We created an example case to validate the practicality of VAO Checker, and conducted a usability test to evaluate the effectiveness of this tool. The result showed that it is a useful system for pipeline designers and engineers. It considered the pipeline accessibility within multi-pipes and enhanced the spatial comprehension. The system can be further integrated into BIM software as an API, extended to pipe assembly planning areas, or even referenced for future optimization.

AUTHORS CONTRIBUTIONS

CH developed the methodology and the mathematical models, programmed VAO Checker, carried out the usability test, analyzed the results and drafted the manuscript. MH assisted the literature review and the usability test. SC offered suggestion and guidance to the research. All authors read and approved the final manuscript.

ACKNOWLEDGEMENTS

This research was supported in part by Taiwan¿s Ministry of Economic Affairs, under contract 101-EC-17-A-15-S1-223. The authors are grateful to Mr. Ching-Yang Kao and Mr. Ming-Fa Lin of United Microelectronics Corporation (UMC), Mr. Yuan-Fu Liao and Mr. Yi-Ti Tsai of CTCI Corporation, Mr. Chien-Chih Lai of L&K Engineering Corporation and Mr. Ching-Hsien Lee of Research Center for BIM, National Taiwan University for their kind support and assistance in this research. We like to express our appreciation to the interviewees of these corporations and organizations.

REFERENCES

1.   Biehl, WH, & Inman, JA (2010). Energy optimization for water systems. *Journal of American Water Works Association, 102*, 6.

2.   Calixto, EES, Bordeira, PG, Calazans, HT, Tavares, CAC, Rodriguez, MTD (2009). Plant design project automation using an automatic pipe routing routine. *Computer Aided Chemical Engineering, 27*, 807–812.doi: 10.1016/S1570-7946(09)70355-4

3.   Chang, HS, Kang, SC, Chen, PH (2009). Systematic procedure of determining an ideal color scheme on 4D models. *Advanced Engineering Informatics, 23*(4), 463–473.doi: 10.1016/j. aei.2009.05.002

4.   Chen, YH, Tsai, MH, Kang, SC, Liu, CW (2013). Selection and evaluation of color scheme for 4D construction models. *Journal of Information Technology in Construction, 18*, 1–19.

5.   Deliang, L, & Huibiao, L (2009). Interfere-check applying to 3D automatic pipe route arrangement.*Proceedings of International Conference on Computational Intelligence and Software Engineering, Wuhan, ?*, 11–13.doi:10.1109/cise.2009.5365920

6.   Feng, H, Fu, Y, Li, L (2012). Layout space modeling for automation design of pipeline system.*Proceedings of 2012 International Conference on Mechatronics and Automation (ICMA), Chengdu, ?*, 5–8.doi:10.1109/icma.2012.6283259

7.   Gao, Z, Walters, RC, Jaselskis, EJ, Wipf, TJ (2006). Approaches to improving the quality of construction drawings from owner's perspective. *Journal of Construction Engineering and Management, 132*(11), 1187–1192.doi: 10.1061/(asce)0733-9364(2006)132:11(1187)

8.   Grootjans, R (2009). XNA 3.0 Game Programming Recipes: A Problem-Solution Approach. Apress, New York.

9.   Guirardello, R, & Swaney, RE (2005). Optimization of process plant layout with pipe routing.*Computers and Chemical Engineering, 30*(1), 99–114.10.1016/j.compehemeng.2005.08.009

10. Ito, T (1999). A genetic algorithm approach to piping route path planning. *Journal of Intelligent Manufacturing, 10*(1), 103–114.10.1023/a:1008924832167

11. Junnila, S, Horvath, A, Guggemos, AA (2006). Life-cycle assessment of office buildings in Europe and the United States. *Journal of Infrastructure Systems, 12*(1), 10–17.doi:10.1061/(asce)1076-0342(2006)12:1(10)

12. Khanzode, A, Fischer, M, Reed, D (2008). Benefits and lessons learned of implementing building virtual design and construction (VDC) technologies for coordination of mechanical, electrical, and plumbing (MEP) systems on a large healthcare project. *Journal of Information Technology in Construction, 13*, 324–342.

13. Kim, D, Corne, D, Ross, P (1996). Industrial plant pipe-route optimisation with genetic algorithms.*Lecture Notes in Computer Science, 1141*, 1012–1021.

14. Korde, T, Wang, Y, Russell, A (2005). Visualization Of Construction Data. Proceedings of 6th Construction Specialty Conference, Toronto, Canada.

15. Kuo, CH, Tsai, MH, Kang, SC (2011). A framework of information visualization for multi-system construction. *Automation in Construction*, *20*(3), 247–262.

16. Miller, T, & Johnson, D (2010). XNA Game Studio 4.0 Programming: Developing for Windows Phone 7 and Xbox 360. Addison-Wesley Professional, Boston.

17. Mitsuta, T, Kobayashi, Y, Wada, Y, Kiguchi, T, Yoshinaga, T (1987). A knowledge-Based Approach To Routing Problems In Industrial Plant Design. Proceedings Of 6th International Workshop on Expert Systems & Their Applications. (pp. 28–30). ?, Avignon, France.

18. Nathan, A (2006). Windows Presentation Foundation Unleashed. Sams Publishing, Indianapolis.

19. Newell, RG (1972). An Interactive Approach To Pipe Routing In Process Plants. Proceedings of IFIP Congress 71, London.

20. Park, JH, & Storch, RL (2002). Pipe-routing algorithm development: case study of a ship engine room design. *Expert Systems with Applications*, *23*(3), 299–309.10.1016/s0957-4174(02)00049-0

21. Qian, X, Ren, T, Wang, CE (2008). A survey of pipe routing design. *Proceedings of 2008 Chinese Control and Decision Conference, Yantai, Shandong*, ?, ?.doi:10.1109/ccdc.2008.4598081

22. Riley, DR, Varadan, P, James, JS, Thomas, HR (2005). Benefit-cost metrics for design coordination of mechanical, electrical, and plumbing systems in multistory buildings. *Journal of Construction Engineering and Management*, *131*(8), 877–889.doi: 10.1061/(asce)0733-9364(2005)131:8(877)

23. Rourke, PW (1975). Development of a Three-Dimensional Pipe Routing Algorithm. PhD Dissertation, Lehigh University.

24. Russell, AD, Chiu, CY, Korde, T (2009). Visual representation of construction management data.*Automation in Construction*, *18*(8), 1045–1062.

25. Schmidt-Traub, H, Köster, M, Holtkötter, T, Nipper, N (1998). Conceptual plant layout. *Computers & Chemical Engineering*, *1*, S499–S504.

26. Songer, AD, Hays, B, North, C (2004). Multidimensional visualization of project control data.*Construction Innovation: Information, Process, Management*, *4*(3), 173–190.

27. (2003). Guidance Notes on the Application of Ergonomics to Marine Systems. American Bureau of Shipping, Houston.

28. Tsai, MH, Kang, SC, Hsieh, SH (2010). A three-stage framework for introducing a 4D tool in large consulting firms. *Advanced Engineering Informatics*, *24*(4), 476–489.

29. Tsai, MH, Kang, SC, Hsieh, SH (2013). Lessons learnt from customization of a BIM tool for a design-build company. *Journal of the Chinese Institute of Engineers*, *37*(2), 189–199.

30. Wang, CP (2011). An Approach for Assessing Reachability of Wheelchair Users. ?, National Taiwan University.

31. Wangdahl, GE, Pollock, S, Woodward, JB (1974). Minimum-trajectory pipe routing. *Journal of Ship Research*, *18*(1), 44–49.

32. Zhou, C, & Yin, Y (2010). Pipe assembly planning algorithm by imitating human imaginal thinking.*Assembly Automation*, *30*(1), 66–74.

33. Zhu, D, & Latombe, JC (1991). Mechanization of spatial reasoning for automatic pipe layout design.*Artificial Intelligence for Engineering, Design, Analysis and Manufacturing*, *5*(1), 1–20.

Investigation of Micro- and Nanosized Particle Erosion in a 90° Pipe Bend Using a Two-Phase Discrete Phase Model

M. R. Safaei[1], O. Mahian[2], F. Garoosi[3], K. Hooman[4], A. Karimipour[5], S. N. Kazi[1], and S. Gharehkhani[1]

[1]Department of Mechanical Engineering, Faculty of Engineering, University of Malaya, 50603 Kuala Lumpur, Malaysia

[2]Department of Mechanical Engineering, Faculty of Engineering, Ferdowsi University of Mashhad, Mashhad, Iran

[3]Department of Mechanical Engineering, University of Semnan, Semnan, Iran

[4]School of Mechanical and Mining Engineering, The University of Queensland, St Lucia, Brisbane, QLD 4072, Australia

[5]Department of Mechanical Engineering, Najafabad Branch, Islamic Azad University, Isfahan, Iran

ABSTRACT

This paper addresses erosion prediction in 3-D, 90° elbow for two-phase (solid and liquid) turbulent flow with low volume fraction of copper. For a range of particle sizes from 10 nm to 100 microns and particle volume fractions from 0.00 to 0.04, the simulations were performed for the velocity range of 5–20 m/s. The 3-D governing differential equations were discretized using finite volume method. The influences of size and concentration of micro- and nanoparticles, shear forces, and turbulence on erosion behavior of fluid flow were studied. The model predictions are compared with the earlier studies and a good agreement is found. The results indicate that the erosion rate is directly dependent on particles' size and volume fraction as well as flow velocity. It has been observed that the maximum pressure has direct relationship with the particle volume fraction and velocity but has a reverse relationship with the particle diameter. It also has been noted that there is a threshold velocity as well as a threshold particle size, beyond which significant erosion effects kick in. The average friction factor is independent of the particle size and volume fraction at a given fluid velocity but increases with the increase of inlet velocities.

INTRODUCTION

Erosion-corrosion, defined as the accelerated corrosion following the damage of surface films, is a common cause of failure in a large amount of power plant equipment like pipes, pumps, compressors, vessels, and turbines. It can often be assumed that corrosion is controlled by adjusting the mass transfer while erosion is under the flow of a particulate second phase. This is a credible assumption as corrosion films are brittle-like materials and therefore are eroded easily by impacting particles [1, 2]. This phenomenon has been investigated experimentally in a number of pioneering studies; see [3–7], for instance. Despite recent advances in computational techniques, erosion-corrosion process is yet to be fully resolved with reasonable accuracy. A multitude of reasons for this rather slow development of simulation techniques applied to this problem can be mentioned. For modeling mass transfer near the solid boundaries, it is necessary to solve the governing equations across the mass transfer boundary layer. In aqueous flows this layer may be an order of magnitude shorter than the viscous sublayer. This requires fine meshes in the near-wall region. Utilizing fine near-wall grids with the support of appropriate near-wall turbulence models, the required mass transfer data for corrosive species can be evaluated [8].

Chen et al. [9] studied erosion prediction approach and its usage in oilfield fittings, especially 3-D elbows and plugged tees, using CFX which is a commercially available CFD package. They used RNG - turbulence model along with DPM to track the particles. The results demonstrated that particle rebound and erosion profile have the most significant roles in particles motion inside oilfield geometries. The comparisons also indicated that CFD predictions for erosion are in good agreement with experimental data.

An erosion prediction approach for specifying wear profiles for a 2-D jet impingement test has been developed by Gnanavelu et al. [10]. This prediction model was according to material wear data achieved from laboratory experiments and CFD modeling. They found an appropriate relationship between predicted and experimental data. Although they found that due to some assumptions about particle size and shape, material hardening, numerical errors, and so forth, some essential errors always exist in the calculation.

Mohyaldin et al. [11] have used three methods (empirical, semiempirical, and computational fluid dynamics, i.e., CFD) to model 2-D sand erosion in a pipe, a problem with significant practical application in oil and gas industry. The results of this study have shown that the direct impingement model (semiempirical model) agrees with the results achieved from the discrete phase model (DPM) implemented in CFD whereas the CFD results dramatically underpredict the empirical ones.

Particles, in an erosion problem, can be external to fluid flow; that is, they may be removals from the walls or upstream flow processes. There are, on the other hand, cases where particles are internal to flow like nanofluids. Nanofluids are synthesized by adding highly conductive solid materials to the base fluid, such as water, ethylene glycol, and oil, all with relatively lower thermal conductivity, usually to improve the heat transfer performance of the mixture (compared to that of the base fluid) [12–14]. The idea of adding microparticles to base fluids was presented decades ago; however, microsized particles have the tendency to settle in the suspension, thereby potentially leading to adverse effects. Use of nanofluids, with nanosized particles suspended in the base fluids, would mitigate the issues of fouling and pipe blockings. In addition, the presence of microsized abrasive solid materials will cause erosion and corrosion of pipes and damage pumps and other devices [15].

Routbort et al. [16] have investigated the effect of nanoparticles on erosion in a car radiator. The nanofluids in their study were 1–4% (volume) silicon carbide (SiC) in water as well as 0.1–0.8% (volume) cupric oxide (CuO) in ethylene glycol. Experiments were conducted in the range of 4 m/s–10 m/s (for velocities) and at 90°–30° impact angles. The radiator was made of Al3003 typical radiator material. In their tests, they did not observe any erosion using nanofluids. Just in one case (Cu/water nanofluid, velocity of 9.6 m/s and impact angle of 90°) the galvanic pitting (and not erosion) was observed. In this case, the material loss rate due to galvanic pitting was around 4×10^{-2} µm/hr which indicated that the erosion had the least effect.

In a subsequent study, Routbort et al. [17] have studied the erosion of nanofluids on impeller of a cast aluminum car cooling system. They used 0.1–0.8% (volume) CuO in ethylene glycol and 0.5–4.0% (volume) SiC in water and in ethylene glycol/water (50%-50%) mixture as nanofluids. The experiments were conducted in the range of 2 m/s–10 m/s (for velocities) and at 90°–30° impact angles. The impeller was made of Al3003 material. Their study has shown no weight loss measured after testing 2% (volume), 170 nm SiC/water for more than 700 hours at 8 m/s velocity, that is, no damage to the impellor of a commercial automobile water pump.

However, in their latest report, Routbort et al. [18] have found 0.65% erosion of impeller after hundreds of hours of pumping SiC/water and SiC/ethylene glycol-water (50/50 vol.%) nanofluids at high mass flow rates (20–28 L/min).

In view of the above, comprehensive analysis of nanofluids as erosive materials is yet missing in the literature [19, 20]. In particular, erosion of nanofluids in turbulent flow regime inside industrial fittings is not fully understood. Hence, the present study aims at investigating turbulent flow of dilute water/Cu micro- and nanofluids in a 3-D 90° elbow using finite volume method with standard K- turbulence and DPM. The simulation results for microsized particle flow regime are compared with those in the literature for validation purpose. Special attention was paid to micro- and nanosized copper particles of different solid volume fractions and Reynolds numbers in a commercial elbow.

GOVERNING EQUATIONS OF TURBULENT MICRO- AND NANOFLUIDS EROSION

The underlying physical assumption in this study is that the particles are carried by the flowing fluid. Therefore, continuity, momentum, DPM, and turbulent equations are used to analyze the flow. The spherical particles' velocity is assumed to be the same as those of flowing fluid. Assuming constant thermophysical properties for fluid and particles, the governing equations are as follows [21–23].

Continuity equation:

$$\frac{\partial}{\partial t}\left(\rho\right) + \nabla \cdot \left(\rho \vec{V}\right) = 0. \tag{1}$$

Momentum equation:

$$\frac{\partial}{\partial t}\left(\rho \vec{V}\right) + \nabla \cdot \left(\rho \vec{V}\vec{V}\right) = -\nabla P + \nabla \cdot \left[\mu\left(\nabla \vec{V} + \nabla \vec{V}^T\right)\right] + \rho g. \tag{2}$$

Standard K-ε turbulence model is as follows.

Turbulent kinetic energy transport equation:

$$\frac{\partial\left(\rho k\right)}{\partial t} + \nabla \cdot \left(\rho \vec{V} k\right) = \nabla \cdot \left[\left(\mu + \frac{\mu_t}{\sigma_k}\right)\nabla k\right] + G_k - \rho \varepsilon. \tag{3}$$

Dissipation of turbulent kinetic energy transport equation:

$$\frac{\partial\left(\rho \varepsilon\right)}{\partial t} + \nabla \cdot \left(\rho \vec{V}\varepsilon\right)$$

$$= \nabla \cdot \left[\left(\mu + \frac{\mu_t}{\sigma_\varepsilon}\right)\nabla\varepsilon\right] + \frac{\varepsilon}{k}\left(C_{\varepsilon 1}G_k - \rho\varepsilon C_{\varepsilon 2}\right). \tag{4}$$

The turbulent eddy viscosity obtained from Prandtl-Kolmogorov relation:

$$\mu_t = C_\mu \rho \frac{k^2}{\varepsilon}.$$

(5)

The turbulence kinetic energy production of the mean velocity gradients, G_k, is given as:

$$G_k = \mu_t \nabla \vec{V} \cdot \left(\nabla \vec{V} + \nabla \vec{V}^T \right) - \frac{2}{3} \nabla \cdot \vec{V} \left(3\mu_t \nabla \cdot \vec{V} + \rho k \right).$$

(6)

The constants for the standard - turbulence model in the above formula are represented in Table 1 [24, 25].

Table 1: Coefficients for standard K- turbulent model

C_μ	σ_k	σ_ε	$C_{\varepsilon 1}$	$C_{\varepsilon 2}$
0.09	1	1.3	1.44	1.92

DPM is as follows:

$$m_p \frac{d\vec{v}_p}{dt} = \sum \vec{F},$$

(7)

Where $\vec{F}$ is an external force acting on the particles which for fine particles with high density ratio (more than one) are drag and buoyancy forces [26].

Therefore, the equation of motion can be simplified to the following form:

$$\frac{d\vec{v}_p}{dt} = F_D \left(\vec{v} - \vec{v}_p \right) + \frac{g \left(\rho_p - \rho \right)}{\rho_g},$$

(8)

Where [27]

$$F_D = \left(\frac{18\mu}{\rho_p d_p^2} \right) \left(\frac{C_D \mathrm{Re}_p}{24} \right),$$

(9)

Where in Re_p is the particle Reynolds number and is given as [28–30]

$$\mathrm{Re}_p = \left(\frac{\rho d_p \left| \vec{v}_p - \vec{v} \right|}{\mu} \right).$$

(10)

The drag coefficient, C_D, as a function of the particle Reynolds number is defined by [31, 32]

$$C_D = \frac{24}{\mathrm{Re}} \left(1 + 11.2355 \mathrm{Re}^{0.653} \right) + \frac{(-0.8271)\,\mathrm{Re}}{8.8798 + \mathrm{Re}}.$$

(11)

The solid particle erosion rates are defined as [33, 34]

$$R_{\mathrm{erosion}} = \sum_{p=1}^{N} \left(\frac{\dot{m}_p C\left(d_p\right) f\left(\alpha\right) v^{b(v)}}{A_f} \right),$$

(12)

Where $C\,(d_p)$ is a function of particle diameter, $f(\ )$ is a function of impact angle, α is the angle between the particle trajectory and wall, v is the relative velocity among particles, $b(v)$ is a function of relative velocity among particles, and A_f is the cell face area at the wall [33].

BOUNDARY CONDITIONS

Figure 1 illustrates the schematic of the problem which is analyzed in the present study. The boundary conditions are also indicated in this figure.

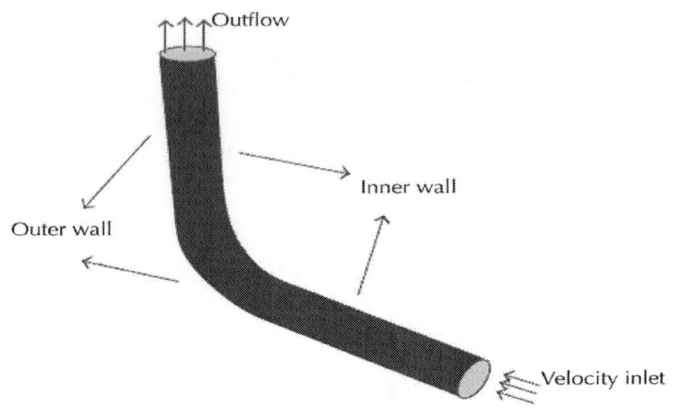

Figure 1: Schematic description of the pipe flow configuration with the elbow being considered for analysis.

NUMERICAL METHOD

The FLUENT commercial code based on finite volume method which has been used in some previous works [21, 22, 35–37] was applied to solve the Reynolds averaged Navier-Stokes (RANS) equations. This method is based on a particular type of the residual weighting approach. In this approach, the computational zone is divided into finite control volumes as each node is covered by a control volume. Eventually, the differential equation is integrated on each finite volume [38–40].

Since in this study the particle volumetric loading ratio is below 10% (0%–4%), the DPM was applied for solving the diluted fluid-solid multiphase flow [41]. As such, the continuous phase, fluid, was simulated by utilizing the Eulerian approach whereas Lagrangian approach was used for modeling the particle phase. Standard wall functions were selected along with standard K- model described above.

The second-order upwind method [42–44] was chosen for the discretization of all terms, while the SIMPLEC algorithm (SIMPLE-Consistent) [15, 45, 46] was employed for pressure-velocity coupling. The impact angle function was specified utilizing a piecewise linear profile as per Table 2. The velocity exponent function and diameter function were fixed at 2.6 and 1.8×10^{-09}, respectively, following [11]. The solution was converged when the residuals for all the equations dropped below 10^{-6} [38, 47].

Table 2: Point values for impact angle function [11]

Point	Angle	Value
1	0	0
2	20	0.8
3	30	1
4	45	0.5
5	90	0.4

NUMERICAL PROCEDURE VALIDATION

Validation with Numerical Study

In order to verify the present simulation, the results from this work were compared with those of [11] where sand erosion in a 2-D elbow was simulated. The geometry was a 50 mm diameter elbow with two 100 mm straight pipes protruded from both sides. The two-phase (air/sand) dilute slurry flow with sand as the dispersed phase was injected at 0.000886 kg/s to the continuous phase, here air, with an inlet velocity of 20 m/s. The variations of total erosion rate and maximum erosion rate with velocity were compared with the results reported by Mohyaldin et al. [11], as shown in Figures 2(a) and 2(b), to observe an excellent agreement between the results.

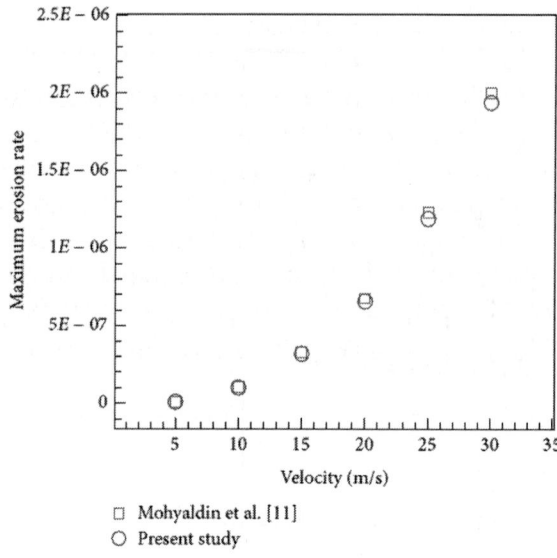

(a) Variation of maximum erosion rate with velocity

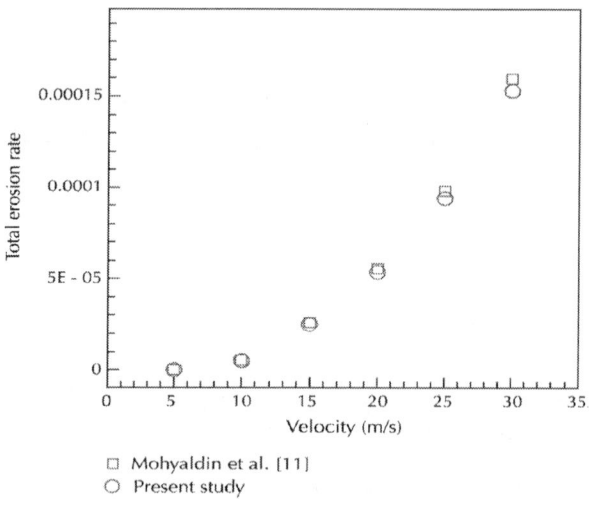

(b) Variation of total erosion rate with velocity

Figure 2: Comparison of total erosion rate and maximum erosion rate predicted here with those of [11].

Validation with Experimental Study

The numerical predictions based on our work were also compared with numerical and experimental results reported by Chen et al. [9] for erosion in elbows and plugged tees. Comparisons were performed for a 2.54 cm (diameter) elbow with a curvature ratio of 1.5 where sand particles of 150-micron diameter are injected at 2.08×10^{-4} kg/s over a range of air/sand velocities: 15.24, 30.48, and 45.72 m/s. The computed average mass loss for elbow was successfully compared with measurements reported in Chen et al. [9], as shown in Figure 3.

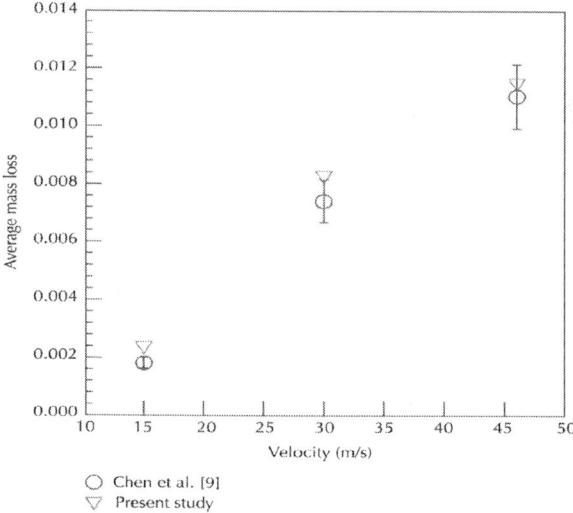

Figure 3: Comparison of average mass loss variations with previous work.

GRID INDEPENDENCE

The computational zone was discretized through structured, nonuniform hexahedral grid distributions. The refined grid was used at the vicinity of the walls where sharp gradients are expected. Several grid distributions were examined as Table 3 indicates. As seen, the effect of grid refinement beyond 61440 grids on the average erosion rate is insignificant implying grid independence of our results.

Table 3: Grid independence tests

Number of grids (V = 20 m/s, φ = 2%)	30720	61440	122880
Average erosion rate for 100 µm particles	6.9523×10^{-6}	6.7833×10^{-6}	6.6965×10^{-6}
Number of grids (V = 20 m/s, φ = 2%)	30720	61440	122880
Average erosion rate for 10 nm particles	2.6789×10^{-6}	2.5029×10^{-6}	2.4351×10^{-6}
Number of grids (V = 20 m/s, φ = 4%)	30720	61440	122880
Average erosion rate for 100 µm particles	1.5270×10^{-5}	1.3857×10^{-5}	1.2994×10^{-5}
Number of grids (V = 20 m/s, φ = 4%)	30720	61440	122880
Average erosion rate for 10 nm particles	4.3001×10^{-6}	4.1646×10^{-6}	4.0843×10^{-6}

RESULTS AND DISCUSSION

In this work, the turbulent fluid flow of water and copper micro- and nanoparticle suspensions through a 90° elbow has been investigated. The material of the 0.0032 m (1/8 inches) diameter elbow was aluminum (3003 Alloy). The length of the two attached pipe pieces at the beginning and the end of the elbow was 0.016 m (5/8 inches) long (5 times pipe diameter). The ratio of the bend radius to pipe inside diameter is equal to 1.5. Water was allowed to flow through the pipe at different velocities (5 m/s, 10 m/s, and 20 m/s). It was assumed that the solid particles are spherical and flow at the same velocity as that of water. Different particle diameters (10, 50, and 100 microns as well as 10, 50, and 100 nanometers) and particle volume fractions (2% and

4%) in the suspension were examined.

The Influence of Velocity on Erosion Rate

To investigate the impact of velocity on the maximum erosion rate and total erosion rate, several inlet velocities were simulated. The impact of inlet flow velocity on the total erosion rate is demonstrated in Figures4 (a) and 4(b) for different particle sizes. One notes that the total erosion rates are near zero for inlet velocity less than 5 m/s and particle volume fraction of 2%. For volume fraction of 4%, this quantity is still negligible for inlet velocity less than 5 m/s and particle diameters below 10 microns. This inlet velocity value of 5 m/s can be considered as a "threshold limit" for total erosion rate beyond which the total erosion rate rockets up with an increase in the inlet flow velocity for each particle diameter. These figures also indicate that, with the increase of particle volume fraction, the total erosion rate increases. The maximum of this erosion increase for $\varphi = 4\%$. Is around 4.9 times at V=20m/s and d_p=100 microns, compared to that of $\varphi = 2\%$.

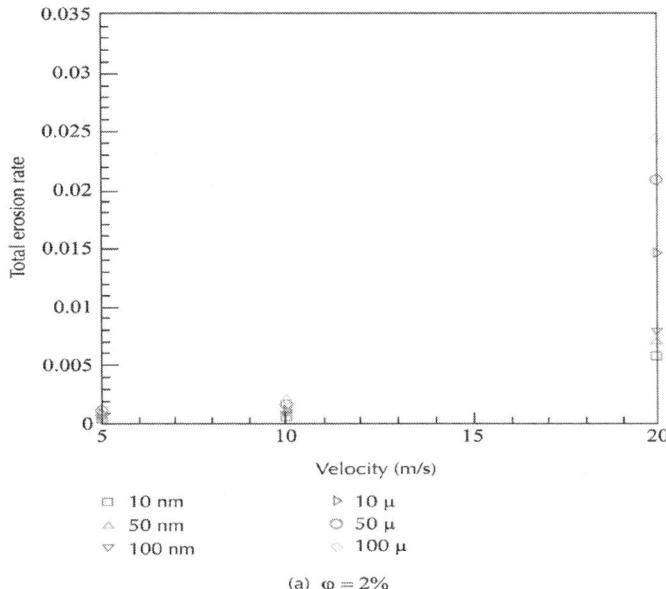

(a) $\varphi = 2\%$

(a) $\varphi = 2\%$

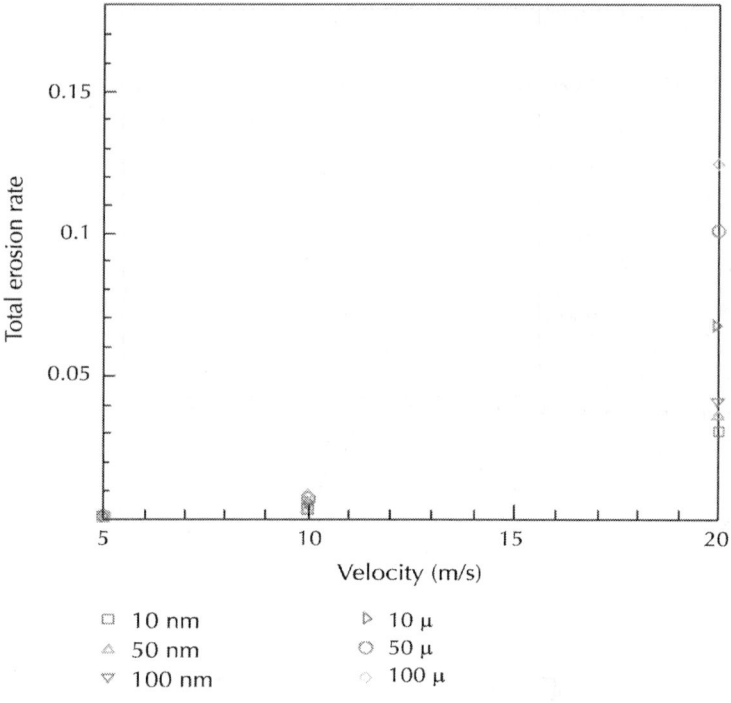

(b) $\varphi = 4\%$

Figure 4: The variation of total erosion rate with velocity.

Similar trends are observed in Figures 5(a) and 5(b) for the maximum erosion rates at six various particle diameters. As seen, the maximum erosion rate is amplified with the particle diameter and velocity increment. This augmentation is negligible at velocities less than 5 m/s, but the difference between the values is more pronounced with an increase in the inlet velocity. Thus, when velocity is increased from 10 m/s to 20 m/s, the maximum erosion rate increases by about an order of magnitude, in fact, by around 7.5 times and 9 times at $\varphi = 2\%$ and $\varphi = 4\%$, respectively.

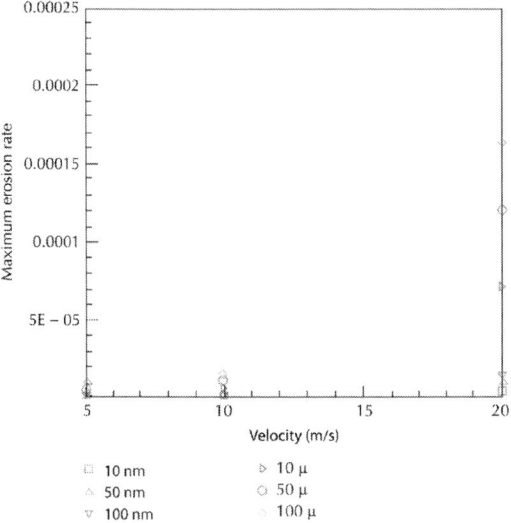

(a) $\varphi = 2\%$

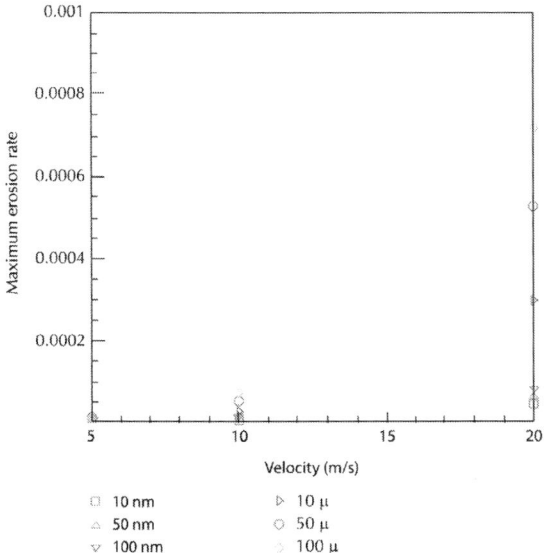

(b) $\varphi = 4\%$

Figure 5: The variation of maximum erosion rate with velocity.

The Effect of Particle Dimension on Erosion Rate

It is significant to study the effect of particle diameter on fluid-solid interaction as particles' size in different systems varies to a large extent from nanometer to centimeter. The particle diameter has direct influence on the drag force and, therefore, affects the flow behavior. The influence of particle diameter on maximum erosion rate, total erosion, pressure drop, and friction factor was studied by changing the particle diameter from 10 nm to 100 μm.

The influence of particle size on the maximum erosion rate was represented in Figures 6(a) and 6(b). As seen, the maximum erosion rate is closely related to the fluid velocity where a threshold velocity as well as a threshold particle size can be identified below which erosion is negligible. These figures also indicate that the rate of erosion augments linearly with particle diameter. One also notes that increasing the volume fraction of the particles, with other parameters fixed, will cause higher maximum erosion rate. The average of this increment is around 4.5 times.

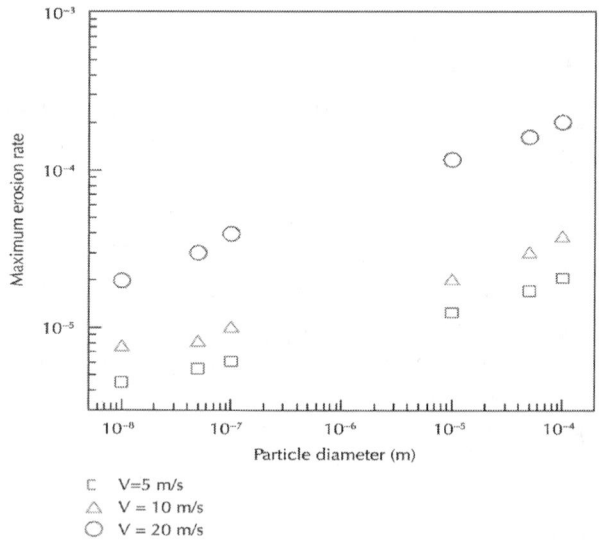

(a) $\varphi = 2\%$

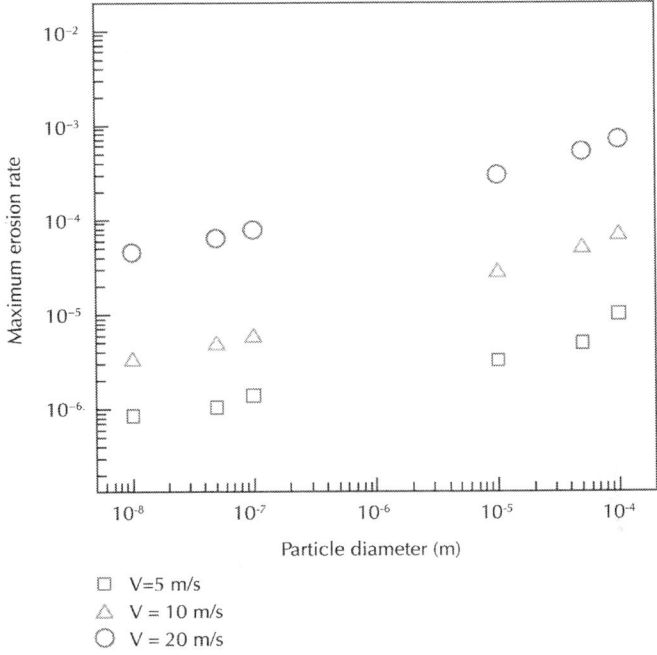

(b) $\varphi = 4\%$

Figure 6: The variation of particle size with maximum erosion rate.

Similar trends are observed in Figures 7(a) and 7(b) for total erosion rate where higher erosion rate is observed when the particle diameter and inlet fluid velocity are increased. This is expected as the particle impact velocity grows with the increase of the inlet flow velocity and particle size (see (12)). However, our numerical results can be used to quantify this increment. Note that the increase in the total erosion rate is around 8.5 times for the increase of velocity from 10 m/s to 20 m/s at $\varphi = 2\%$ and 9.5 times at $\varphi = 4\%$. The influence of volume fraction enhancement on total erosion rate is also around 8 times when the volume fraction is increased from 2% to 4%.

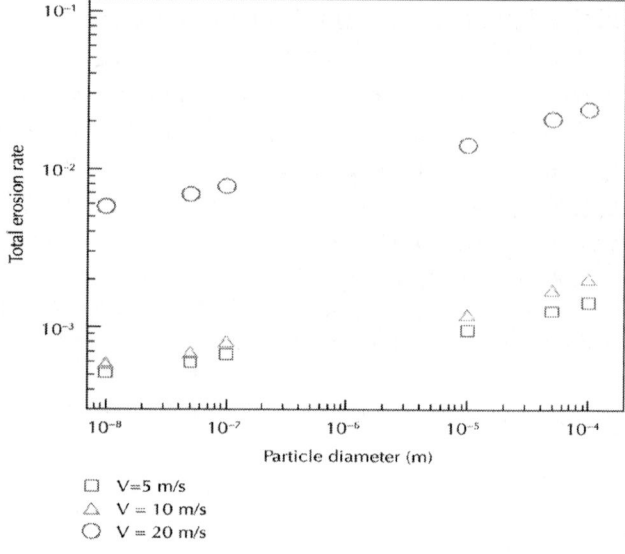

(a) $\varphi = 2\%$

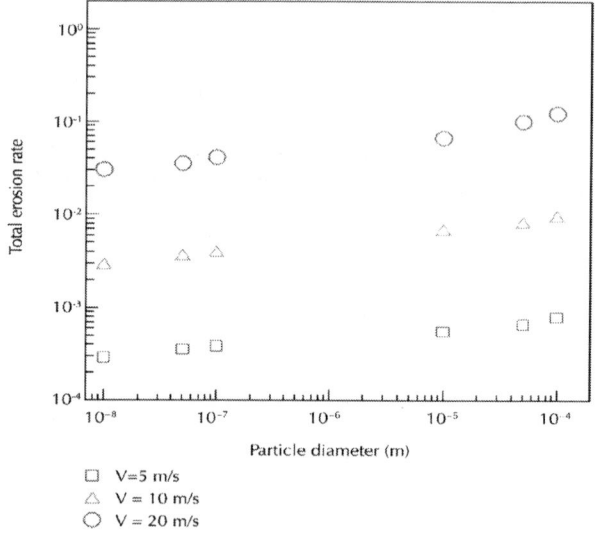

(b) $\varphi = 4\%$

Figure 7: The variation of particle size with total erosion rate.

The declining impact of particle size on the maximum pressure was shown in Figures 8(a) and 8(b). This can be attributed to the reduction in drag forces as a result of an increase in the particle size. Consequently, with the same particle volume fraction, particle numbers are lowered compared to the case with smaller particles. The figures also indicate that there is a direct relationship between the velocity and increase of maximum pressure. It is also clear from the figures that an increase in particle volume fraction leads to higher maximum pressure. As a result, the maximum pressure value is observed when 10 nm particles at 4% volume fraction flow with water at 20 m/s.

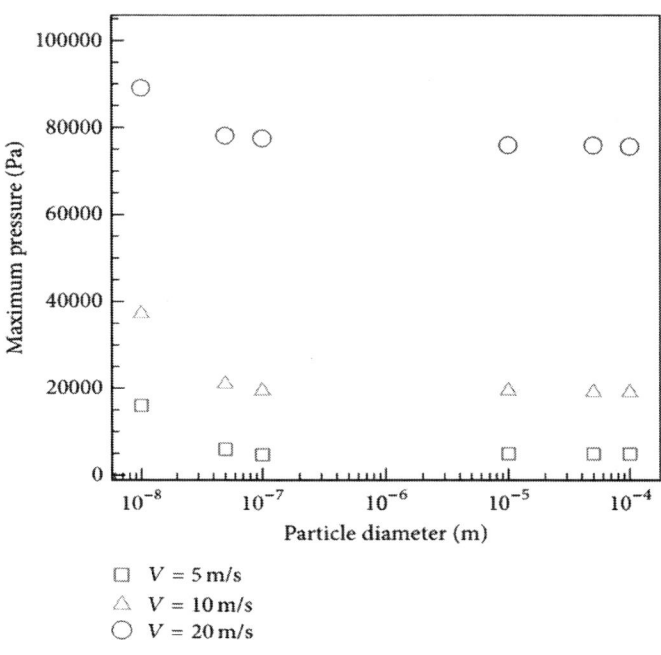

(a) $\varphi = 2\%$

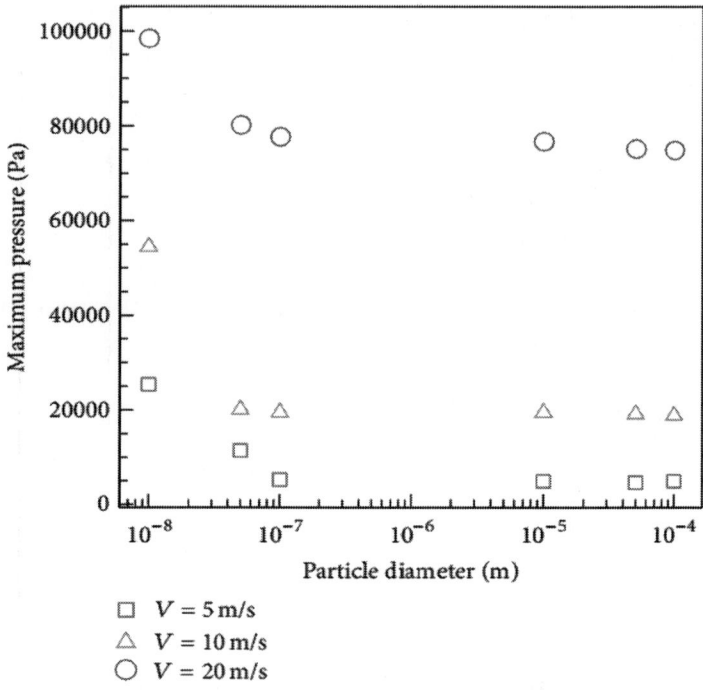

(b) $\varphi = 4\%$

Figure 8: The variation of particle size with maximum pressure.

Interestingly, according to Figures 9(a) and 9(b), the average friction factor—which has been calculated based on Fanning equation—is insensitive to either the particle size or volume fraction. However, one observes that the average friction factor increases with inlet velocity unlike a single-phase flow.

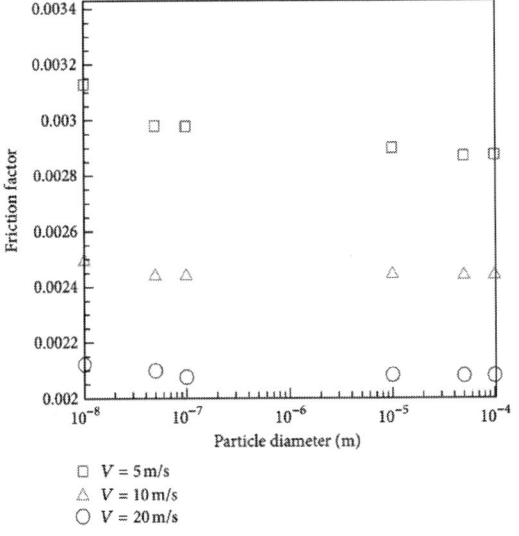

(a) $\varphi = 2\%$

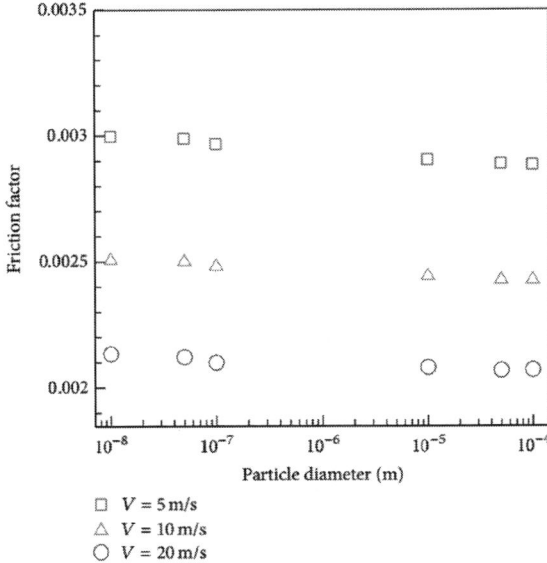

(b) $\varphi = 4\%$

Figure 9: The variation of particle size with average friction factor.

Figure 10 illustrates the erosion contour inside the elbow for V=20 m/s, particle size = 100 microns, and the volume fractions of (Cu) 2%. As seen, the maximum erosion is observed near the midpoint, along the symmetry plane of the pipe bend, which is the location where velocity profiles begin an inverse behavior and the pressure is maximum.

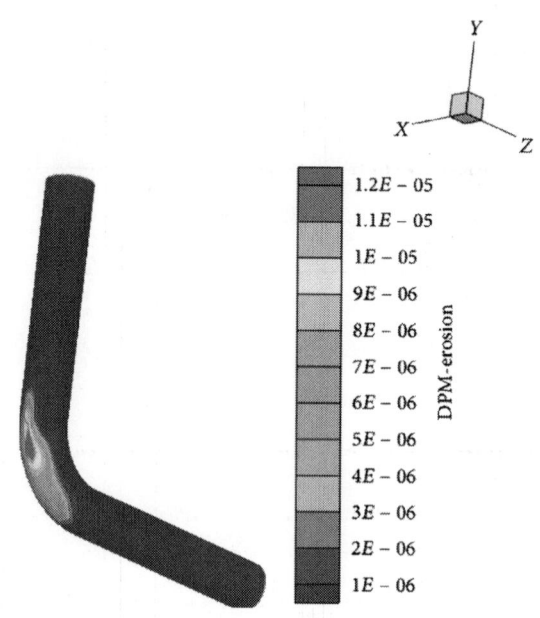

Figure 10: Erosion contour on the wall of the bend.

Finally, for engineering applications and presentation of the physical influence of the parameters, the following single nonlinear correlation is derived from Figures 11(a) and 11(b) to estimate the average erosion rate as a function of particles' concentration, diameter, and inlet velocity, valid for the range of parameters in this work; that is $0.02 \leq \varphi \leq 0.04$, $5\,\text{m/s} \leq V \leq 20\,\text{m/s}$, and $10\,\text{nm} \leq d_p \leq 100$ microns. The average deviation of this correlation is 9.5%. Consider the following:

Average erosion rate (AER)

$$= 3.6667 \times 10^{-8} \left(\varphi^{1.0024} V^{3.4953} d_p^{0.1399} \right).$$

$$(13)$$

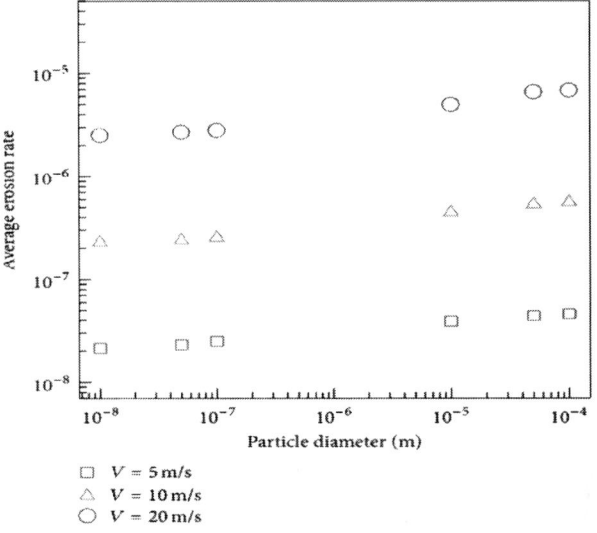

(a) $\varphi = 2\%$

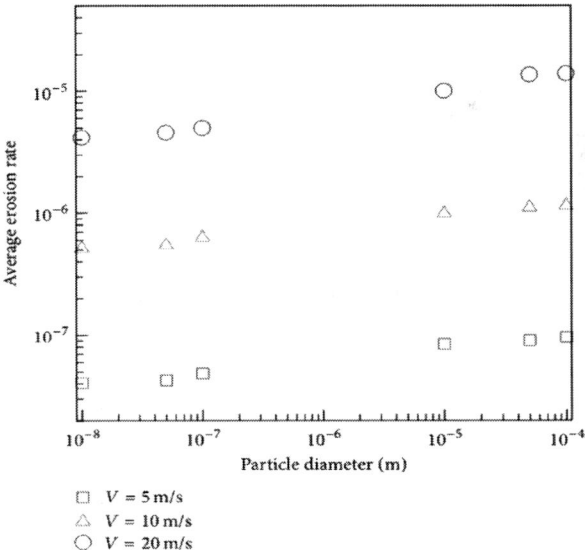

(b) $\varphi = 4\%$

Figure 11: The variation of particle size with average erosion rate.

CONCLUSIONS

A numerical study of erosion in turbulent water-based/copper (Cu) micro- and nanosized fluid flow through a 90° elbow has been conducted. Different solid volume fractions, particle sizes, and velocities were considered along with the maximum erosion rate, total erosion rate, average erosion rate, friction factor, and maximum pressure.

The conclusions are summarized as follows.

- There is a threshold velocity as well as a threshold particle size, beyond which erosion is significant.
- The maximum erosion rate, average erosion rate, and total erosion rate increase with particle diameter, volume fraction, and inlet fluid velocity.
- Increase of the particle diameter decreases the maximum pressure.
- An increase in particle volume fraction or velocity augments the maximum pressure.
- The average friction factor does not depend on particle size and/ or volume fraction for a given flow rate.
- With the increase of the inlet velocity, the average friction factor enhances.

The usage of nanofluids in heat transfer has an obvious benefit from the thermal efficiency point of view. Nonetheless, care must be taken as depending on particle size, fluid velocity, particle shape, particle sedimentation, particle agglomeration, and surface erosion adverse effects can negate the benefits associated with heat transfer augmentation.

ACKNOWLEDGMENTS

The authors gratefully acknowledge the High Impact Research Grant UM.C/HIR/MOHE/ENG/45, UMRG Grant RP012D-13AET, and Faculty of Engineering, University of Malaya, Malaysia, for support in conducting this research work.

REFERENCES

1. A. Keating and S. Nesic, "Prediction of two-phase erosion-corrosion in bends," in Proceedings of the 2nd International Conference on CFD in the Minerals and Process Industries (CSIRO '99), pp. 229–236, Melbourne, Austrlia, 1999.

2. S. Nešić, "Using computational fluid dynamics in combating erosion-corrosion," Chemical Engineering Science, vol. 61, no. 12, pp. 4086–4097, 2006.

3. J. Southard, R. A. Young, and C. D. Hollister, "Experimental erosion of calcareous ooze," Journal of Geophysical Research, vol. 76, no. 24, pp. 5903–5909, 1971.

4. P. Lonsdale and J. B. Southard, "Experimental erosion of North Pacific red clay," Marine Geology, vol. 17, no. 1, pp. M51–M60, 1974.

5. M. E. Gulden, "Correlation of experimental erosion data with elastic—plastic impact models," Journal of the American Ceramic Society, vol. 64, no. 3, pp. C59–C60, 1981.

6. R. A. Saravanan, M. K. Surappa, and B. N. Pramila Bai, "Erosion of A356 Al-SiCp composites due to multiple particle impact," Wear, vol. 202, no. 2, pp. 154–164, 1997.

7. G. T. Burstein and K. Sasaki, "Effect of impact angle on the slurry erosion-corrosion of 304L stainless steel," Wear vol. 240, no. 1-2, pp. 80–94, 2000.

8. W. H. Ahmed, M. M. Bello, M. El Nakla, and A. Al Sarkhi, "Flow and mass transfer downstream of an orifice under flow accelerated corrosion conditions," Nuclear Engineering and Design, vol. 252, pp. 52–67, 2012.

9. X. Chen, B. S. McLaury, and S. A. Shirazi, "Application and experimental validation of a computational fluid dynamics (CFD)-based erosion prediction model in elbows and plugged tees," Computers and Fluids, vol. 33, no. 10, pp. 1251–1272, 2004.

10. A. Gnanavelu, N. Kapur, A. Neville, J. F. Flores, and N. Ghorbani, "A numerical investigation of a geometry independent integrated method to predict erosion rates in slurry erosion," Wear vol. 271, no. 5- 6, pp. 712–719, 2011.

11. M. E. Mohyaldin, N. Elkhatib, and M. C. Ismail, "Evaluation of different modelling methods used for erosion prediction," in Proceedings of the NACE Corrosion Changhai Conference & Expo, pp. 1–19, Changhai, China, 2011.

12. M. H. Esfe, M. Akbari, D. Toghraie, A. Karimipour, and M. Afrand, "Effect of nanofluid variable properties on mixed convection flow and heat transfer in an inclined two-sided lid-driven cavity with sinusoidal heating on sidewalls," Heat Transfer Research, vol. 45, no. 5, pp. 409–432, 2014.

13. M. R. Safaei, H. Togun, K. Vafai, S. N. Kazi, and A. Badarudin, "Investigation of thermal conductivity and rheological properties of nanofluids containing graphene nanoplatelets," Numerical Heat Transfer A, vol. 66, no. 12, pp. 1321–1340, 2014.

14. H. Togun, M. R. Safaei, R. Sadri et al., "Numerical simulation of laminar to turbulent nanofluid flow and heat transfer over a backward-facing step," Applied Mathematics and Computation, vol. 239, pp. 153–170, 2014.

15. M. Goodarzi, M. R. Safaei, K. Vafai et al., "Investigation of nanofluid mixed convection in a shallow cavity using a two-phase mixture model," International Journal of Thermal Sciences, vol. 75, pp. 204–220, 2014.

16. J. Routbort, D. Singh, W. Yu et al., "Effects of nanofluids on heavy vehicle cooling systems," inProceedings of the VT Annual Merit Review Meeting, pp. 1–16, Argonne National Laboratory, February 2008.

17. J. Routbort, D. Singh, E. Timofeeva, W. Yu, and R. Smith, Erosion of Radiator Materials by Nanofluids, Argonne National Laboratory, Vehicle Technologies–Annual Review, 2010.

18. J. Routbort, D. Singh, E. Timofeeva, W. Yu, and R. Smith, Erosion of Radiator Materialsby Nanofluids, Vehicle Technologies— Annual Review, Argona National Lab., 2011.

19. A. Karimipour, M. H. Esfe, M. R. Safaei, D. T. Semiromi, S. Jafari, and S. N. Kazi, "Mixed convection of copper-water nanofluid in a shallow inclined lid driven cavity using the lattice Boltzmann method,"Physica A: Statistical Mechanics and its Applications, vol. 402, pp. 150–168, 2014.

20. M. H. Esfe, S. S. M. Esforjani, M. Akbari, and A. Karimipour, "Mixed-convection flow in a lid-driven square cavity filled with a nanofluid with variable properties: effect of the nanoparticle diameter and of the position of a hot obstacle," Heat Transfer Research, vol. 45, no. 6, pp. 563–578, 2014.

21. H. Badr, M. Habib, R. Ben-Mansour, and S. Said, "Effect of flow velocity and particle size on erosion in a pipe with sudden contraction," in Proceedings of the 6th Saudi Engineering Conference, vol. 5, pp. 79–88, KFUPM, Dhahran, Saudi Arabia, December 2002.

22. H. M. Badr, M. A. Habib, R. Ben-Mansour, and S. A. M. Said, "Numerical investigation of erosion threshold velocity in a pipe with sudden contraction," Computers and Fluids, vol. 34, no. 6, pp. 721–742, 2005.

23. Q. H. Mazumder, "Effect of liquid and gas velocities on magnitude and location of maximum erosion in U-bend," Open Journal of Fluid Dynamics, vol. 2, pp. 29–34, 2012.

24. M. R. Safaei, H. R. Goshayeshi, B. S. Razavi, and M. Goodarzi, "Numerical investigation of laminar and turbulent mixed convection in a shallow water-filled enclosure by various turbulence methods,"Scientific Research and Essays, vol. 6, no. 22, pp. 4826–4838, 2011.

25. M. R. Safaei, Y. Maghmoumi, and A. Karimipour, "Numerical investigation of turbulence mixed convection heat transfer of water and drilling mud inside a square enclosure by finite volume method," in Proceedings of the International Meeting on Advances inThermofluids (IMAT ‹11), Melaka, Malaysia, October 2011.

26. E. Sistani, "PIV measurements around a rotating single gear partially submerged in oil within modelled SAAB gearbox," in Diploma Work-Department of Applied Mechanics, Chalmers University of Technology, Göteborg, Sweden, 2010.

27. M. A. Habib, R. Ben-Mansour, H. M. Badr, and M. E. Kabir, "Erosion and penetration rates of a pipe protruded in a sudden contraction," Computers & Fluids, vol. 37, no. 2, pp. 146–160, 2008.

28. V. Abdolkarimi and R. Mohammadikhah, "CFD modeling of particulates erosive effect on a commercial scale pipeline bend,"

ISRN Chemical Engineering, vol. 2013, Article ID 105912, 10 pages, 2013.

29. H. Zhu, H. Zhao, Q. Pan, and X. Li, "Coupling analysis of fluid-structure interaction and flow erosion of gas-solid flow in elbow pipe," Advances in Mechanical Engineering, vol. 2014, Article ID 815945, 10 pages, 2014.

30. M. G. Lipsett and V. Bhushan, "Modeling erosion wear rates in slurry flotation cells," Journal of Failure Analysis and Prevention, vol. 12, no. 1, pp. 51–65, 2012.

31. M. A. Habib, H. M. Badr, S. A. M. Said, R. Ben-Mansour, and S. S. Al-Anizi, "Solid-particle erosion in the tube end of the tube sheet of a shell-and-tube heat exchanger," International Journal for Numerical Methods in Fluids, vol. 50, no. 8, pp. 885–909, 2006.

32. H. Zhu, Y. Lin, G. Feng et al., "Numerical analysis of flow erosion on sand discharge pipe in nitrogen drilling," Advances in Mechanical Engineering, vol. 2013, Article ID 952652, 10 pages, 2013.

33. A. Campos-Amezcua, A. Gallegos-Muñoz, C. Alejandro Romero, Z. Mazur-Czerwiec, and R. Campos-Amezcua, "Numerical investigation of the solid particle erosion rate in a steam turbine nozzle," Applied Thermal Engineering, vol. 27, no. 14-15, pp. 2394–2403, 2007.

34. P. Frawley, J. Corish, A. Niven, and M. Geron, "Combination of CFD and DOE to analyse solid particle erosion in elbows," International Journal of Computational Fluid Dynamics, vol. 23, no. 5, pp. 411–426, 2009.

35. H. M. Badr, M. A. Habib, R. Ben-Mansour, S. A. M. Said, and S. S. Al-Anizi, "Erosion in the tube entrance region of an air-cooled heat exchanger," International Journal of Impact Engineering, vol. 32, no. 9, pp. 1440–1463, 2006.

36. M. A. Habib, H. M. Badr, R. Ben-Mansour, and M. E. Kabir, "Erosion rate correlations of a pipe protruded in an abrupt pipe contraction," International Journal of Impact Engineering, vol. 34, no. 8, pp. 1350–1369, 2007.

37. K. Sun, L. Lu, and H. Jin, "Modeling and numerical analysis of the solid particle erosion in curved ducts," Abstract and Applied Analysis, vol. 2013, Article ID 245074, 8 pages, 2013.

38. A. Karimipour, M. Afrand, M. Akbari, and M. Safaei, "Simulation of fluid flow and heat transfer in the inclined enclosure," International Journal of Mechanical and Aerospace Engineering, vol. 6, pp. 86–91, 2012.

39. S. V. Patankar, Numerical Heat Transfer and Fluid Flow, Taylor & Francis, 1980.

40. M. R. Safaei, B. Rahmanian, and M. Goodarzi, "Numerical study of laminar mixed convection heat transfer of power-law non-Newtonian fluids in square enclosures by finite volume method,"International Journal of Physical Sciences, vol. 6, no. 33, pp. 7456–7470, 2011.

41. P. Xu, Z. Wu, A. S. Mujumdar, and B. Yu, "Innovative hydrocyclone inlet designs to reduce erosion-induced wear in mineral dewatering processes," Drying Technology, vol. 27, no. 2, pp. 201–211, 2009.

42. M. Goodarzi, M. R. Safaei, H. F. Oztop et al., "Numerical study of entropy generation due to coupled laminar and turbulent mixed convection and thermal radiation in an enclosure filled with a semitransparent medium," The Scientific World Journal, vol. 2014, Article ID 761745, 8 pages, 2014.

43. H. Goshayeshi, M. Safaei, and Y. Maghmoumi, "Numerical simulation of unsteady turbulent and laminar mixed convection in rectangular enclosure with hot upper moving wall by finite volume method," in Proceedings of the 6th International Chemical Engineering Congress and Exhibition (ICheC ⟨09), Kish Island, Iran, 2009.

44. M. R. Safaiy, S. R. Saleh, and M. Goudarzi, "Numerical studies of laminar natural convection in a square cavity with orthogonal grid mesh by finite volume method," International Journal of Advanced Design and Manufacturing Technology, vol. 1, no. 2, pp. 13–21, 2011.

45. M. Goodarzi, M. R. Safaei, A. Karimipour et al., "Comparison of the finite volume and lattice Boltzmann methods for solving natural convection heat transfer problems inside cavities and enclosures," Abstract and Applied Analysis, vol. 2014, Article ID 762184, 15 pages, 2014.

46. M. R. Safaiy and H. R. Goshayeshi, "Numerical simulation of laminar and turbulent mixed convection in rectangular enclosure

with hot upper moving wall," International Journal of Advanced Design and Manufacturing Technology, vol. 3, no. 2, pp. 49–57, 2011.

47. M. R. Safaei, M. Goodarzi, and M. Mohammadi, "Numerical modeling of turbulence mixed convection heat transfer in air filled enclosures by finite volume method," The International Journal of Multiphysics, vol. 5, no. 4, pp. 307–324, 2011.

Chapter 7

Studying the Effect of Some Surfactants on Drag Reduction of Crude Oil Flow

Ali A. Abdul-Hadi[1] and Anees A. Khadom[2]

[1]Chemical Engineering Department, College of Engineering, University of Baghdad, Al Jadriya, Baghdad, Iraq

[2]Chemical Engineering Department, College of Engineering, University of Diyala, Baquba 32001, Daiyla, Iraq

ABSTRACT

The influence of SDBS, SLS, SLES, and SS as drag reducing agents on flow of Iraqi crude oil in pipelines was investigated in the present work. The effect of additive type, additive concentration, pipe diameter, solution flow rate, and the presence of elbows on the percentage of drag reduction (%Dr) and the amount of flow increases (%FI) was addressed. The maximum drag reduction was 55% obtained at 250 ppm SDBS surfactant flowing in straight pipes of 0.0508 m I.D. The dimensional analysis was used for grouping the significant quantities into dimensionless groups to reduce the number of variables. The results showed good agreement between the observed drag reduction percent values and the predicted ones with high value of the correlation coefficient.

INTRODUCTION

Drag reduction is a phenomenon in which the friction of a liquid flowing in a pipe in turbulent flow is decreased by using a small amount of an additive. The used drag reducing additives are effective because they reduced the turbulent friction of the solution. This resulted in a decrease in the pressure drop across a length of the pipe and likewise reduced the energy required to transport the liquid [1]. Surfactants are one of the most important drag reducing agents, which have the ability to form a certain structure called micelles. The important aspect of surfactant which impacts their performance is their ability to self-repair. This is the ability of a group of molecules to return to its original form after their structure has been altered as a result of high shear; this property recognizes the surfactant from polymers and aluminum disoaps, which degrade when subjected to high shear and generally cannot reform. Therefore, they cannot be effective in recirculating the fluid, and these pumps apply high shear stress to fluid. This causes the polymer chains to break into small segments which do not have the ability to revert to their original form. On the other hand, surfactants are able to repair themselves in a matter of seconds upon degradation of shear. This characteristic makes surfactants a good candidate for recirculation systems [2]. The mechanisms by which these agents work (turbulent suppression; extension of laminar behavior to abnormally high Reynolds numbers; or wall layer modification, reduction of friction in fully developed turbulence) are not defectively established, but they are believed to inhibit the formation of microscopic eddies in the liquid [3]. The goal of the present work was to investigate the validity of the effectiveness of SDBS, SLS, SLES and SS (concentrations of 50, 100, 150, 200, and 250 ppm) as drag reducing agents with Kirkuk crude oil. Also to study the effect of additive type, additive concentration, pipe diameter, solution flow rate, and the presence of radius elbows on the percentage of drag reduction (%Dr) and the amount of flow increases (%Fl), these parameters have the most significant effect on the flow of fluids.

MECHANISM OF DRAG REDUCTION

The flow in most crude oil pipelines is turbulent. This means that most of the drag, or energy loss while pumping, is due to turbulent eddies in the oil rather than the friction from pipeline walls. Drag reduction agents are chemicals that are injected into a crude oil pipeline to reduce the energy loss; this produces a solution in pressure drop smaller than that which would occur with untreated solvent moving at the same flow rate. Drag reduction agents are described as a thick, viscous liquid with the appearance of old honey and highly viscoelastic. DRAs used in oil and products pipelines are themselves hydrocarbons and thus should have no effect on physical properties of refining processes or refined products. DRA-solvent solutions are viscoelastic, time-independent, shear degradable, and non-Newtonian fluids [4, 5]. Several types of additives cause drag reducing phenomena to occur. Surfactants are the focus of this research because they were used as a drag reduction agent. The surfactants of different types anionic, nonionic, Zwitterionic, and cationic behave in a characteristic manner in solutions. In these solutions, the hydrophobic group avoids contact with polar molecules by forming micelles. In micelles, the hydrophilic parts, which are polar, contact the polar molecules allowing the nonpolar, hydrophobic parts to concentrate in the center of the micelle. Surfactants as well as the structure of the micelle both contribute to the high drag reducing properties of the molecule. An important aspect of drag reducing surfactant additives which impacts their performance is their ability to self-repair. This is the ability of a group of molecules to return to their original form after its structure has been altered as a result of high shear [6].

EXPERIMENTAL WORK

Liquids

Kirkuk crude oil (Kirkuk governorate, Iraq) which was used in the present work (provided from Al-Dura refinery, Iraq). The physical properties of this crude oil were 2.296 viscosities @ 25°C (c.st), 0.8513 specific gravity, and 35.40 API. The kinematic viscosity of Iraqi crude oil was calculated according to ASTM D-445, while specific gravity was according to ASTM D 1217-81.

Surfactants

SDBS, SLS, SLES, and SS are anionic surfactants which were used as drag reducing agents (concentrations 50, 100, 150, 200, and 250 ppm) in the present work. They were supplied by General Company of Vegetable Oil Industries, Baghdad, Iraq. The specifications and some physical properties are shown in Table 1.

Table 1: Specification of surfactants

Surfactants	Scientific name	Chemical structure	Molecular weight
SDBS	Sodium dodecyl-benzene sulfonate	$C_{12}H_{25}C_6H_4SO_3-Na$	348
SLS	Sodium lauryl sulfate	$C_{12}H_{26}O_4S-Na$	289
SLES	Sodium laureth Sulfate	$CH_3(CH_2)_{10}CH_2$ $(OCH_2CH_2)_3OSO_3-$ Na	372
SS	Sodium stearate	$C_{17}H_{35}COO-Na$	306

Description of Circulating Flow Loop System

Figure 1 represents the schematic diagram of flow system apparatus used in the present work, which consists of reservoir tank of solution (0.88 × 0.88 × 0.88 m³ volume), centrifugal pumps (flow rate = 45 m³/hr; Power = 25 hp) which was used to circulate the solution from the reservoir tank through pipes, while another pump (flow rate = 1 m³/hr; Power = 0.5 hp) was connected to the draining exit of the tank, flow meter (12 m³/h maximum flow rate), valves to control the amount and direction of solution flow rate through the system, pressure gauges, and pipes of different inside diameters (0.0508, 0.0254, and 0.0191 m). These pipes are made of commercial carbon steel with relative roughness shown in Table 2.

Table 2: Relative roughness and length of pipes used

Pipe inside diameter, m	Relative roughness, ε/d	Length of pipe with elbows, m	Length of straight pipe, m
0.0508	0.000885	4.656	3
0.0254	0.001770	2.378	3
0.0191	0.002362	1.687	3

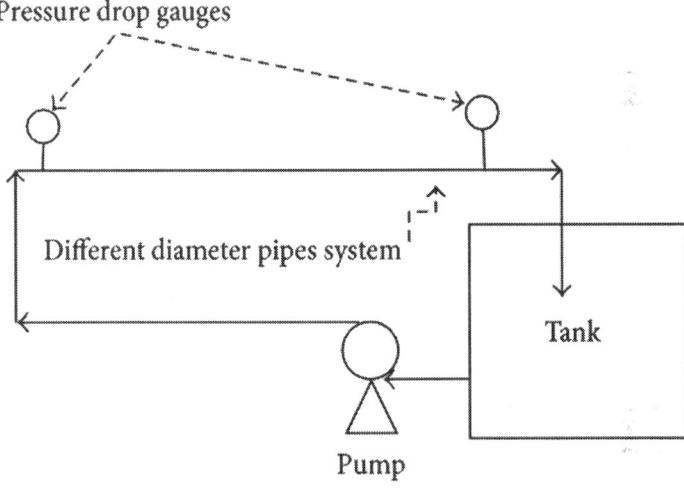

Figure 1: Schematic diagram of flow system.

Experimental Procedure

The preparation of additive solution by mixing small amounts of surfactants with a sample of crude oil is the first step in the experimental procedure; then the solution is added into the reservoir tank of crude oil to use in the recirculation closed system. The operation is started by pumping the solution through the testing section for the same pipe diameter, additive type, and additive concentration. For each run, the flow rate of the solution was controlled bypass section to a certain value, while pressure drop readings were taken. Readings of pressure drop were taken again when the flow rate of the solution was changed to another fixed value. This procedure was repeated for each pipe diameter, additive type, additive concentration, and crudes type.

RESULTS AND DISCUSSION

Results Calculations

Four anionic surfactant types (SDBS, SLS, SLES, and SS) as drag reducing agents with Iraqi crude oil (Kirkuk crude oil) were used. The factorial experimental design was used. The following equations were used to calculate the Reynolds number (Re), percentage drag reduction (%Dr), percentage flow increase (%FI) [7], and friction factor in terms of fanning friction factor [8], respectively:

$$\mathrm{Re} = \frac{\rho \cdot v \cdot d}{\mu},$$

$$\%\mathrm{Dr} = \frac{\Delta P_b - \Delta P_a}{\Delta P_b},$$

$$\%\mathrm{FI} = \left(\frac{1}{1 - (\%\mathrm{Dr}/100)^{0.55}} - 1 \right) \times 100,$$

$$f = \frac{\Delta P \cdot d/4L}{\rho \cdot v^2/2},$$

(1)

where ρ is the density, V is the linear velocity, d is the pipe diameter, μ is the viscosity, ΔP_b and ΔP_a are the pressure drop before and after addition of surfactants, and L is the pipe length.. Table 3 shows the experimental calculation. Similar tables were obtained for other surfactants at different conditions. Table 3 shows the maximum values of %Dr (%Dr$_1$ in pipe with elbows, %Dr$_2$ in straight pipelines) and %FI for all drag reducing agents with Kirkuk crude oil solution. Maximum %Dr$_2$ of 55%, 42%, and 30% were obtained using Kirkuk crude oil containing 250 ppm of SDBS surfactant flowing in straight pipes of 5.08, 2.54, and 1.91 cm I.D., respectively. While the maximum %Dr$_1$

of 48%, 45%, and 32% were obtained using Kirkuk crude oil flowing in the pipes of different lengths (i.e., 1.1 m for 5.08 cm I.D., 0.6 m for 2.54 cm I.D., and 0.35 m for 1.91 cm I.D.), each joined with two elbows of standard radius. The SDBS has a large stability than other additives. This may be attributed to the structure of micelles formed in the surfactant solution and its resistance to the shear forces which governs the effectiveness of the surfactant used as drag reducer. The order of reduction was as follows:

$$SDBS > SLS > SLES > SS. \tag{2}$$

Table 3: Experimental data for 150 ppm SLES surfactant dissolved in the Kirkuk crude oil flowing in 0.0254 m I.D. pipe

Q(m³/hr)	Re	%Dr₁	f₁	%Fl₁
1	6064.57	17.34	0.007561	11.04
2	12129.14	18.25	0.006937	11.72
3	18193.70	19.44	0.006572	12.62
4	24258.27	20.22	0.005930	13.23
5	30322.84	21.18	0.005966	13.99
6	36387.41	22.61	0.005308	15.14
7	42451.97	24.30	0.004962	16.55
8	48516.54	25.33	0.004764	17.43
9	54581.11	26.31	0.004383	18.28
10	60645.68	28.31	0.004115	20.09
11	66710.25	28.67	0.003980	20.42
12	72774.81	30.43	0.003658	22.09

Effect of Surfactant Concentration

Figure 2 shows the effect of surfactant concentration on drag reduction process. The same figures can be obtained at different conditions. These figures show that the %Dr increases with increasing the additive concentration. The increment in %Dr is ascribed to increases of associated additive molecules in the process of drag reduction. Also, it shows that there is no limited value of concentration after which no further drag reduction occurs within additives concentration (50–250 ppm) for surfactants. In order to check that the additives do not affect the physical properties of used crude oil, the viscosity of crude oil was evaluated; the results indicate that there is no change in physical properties after addition. These results agree with the work of Takashi and Hiromoto [9] and others [10].

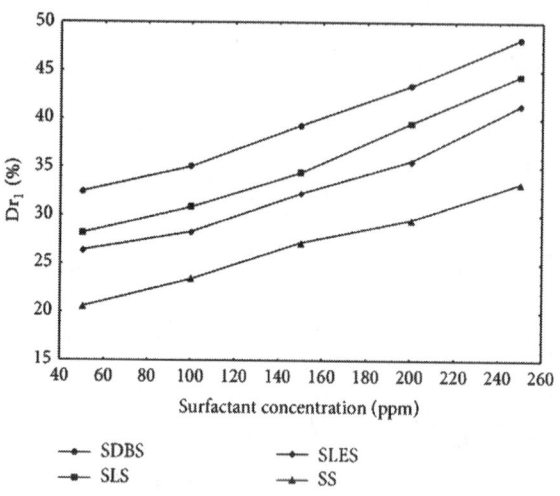

Figure 2: The effect of surfactant concentration on drag reduction for 0.0508 m pipe diameter and 12 m³/h flow rate.

Effect of Pipe Diameter

Figure 3 shows the effect of pipe diameter on %Dr. The comparison of %Dr between the three pipes achieved at a constant flow rate through each of them, certain additive type, and concentration. The results show that %Dr increase with pipe diameter increasing within certain additive type and concentration. This increase in %Dr is attributed to large eddies that exist in the pipe of large diameter, which absorb large amount of energy from the main flow. While in the small pipes, the number of formed small eddies were larger than large eddies formed in the large pipes. These small eddies needed a large amount of energy absorbed from the main flow to overcome the resistance of viscosity and then complete its shape. Not all small eddies absorb equal amount of energy, some of them absorb amount of energy not that is able to overcome viscous resistance and then eventually disappear causing loss in the energy of the main flow, while the other eddies absorb enough energy and enable to overcome viscosity resistance. The %Dr in small pipes is lower than in large pipes due to small eddies which absorb small amount of energy that does not enable it to overcome viscosity resistance [11].

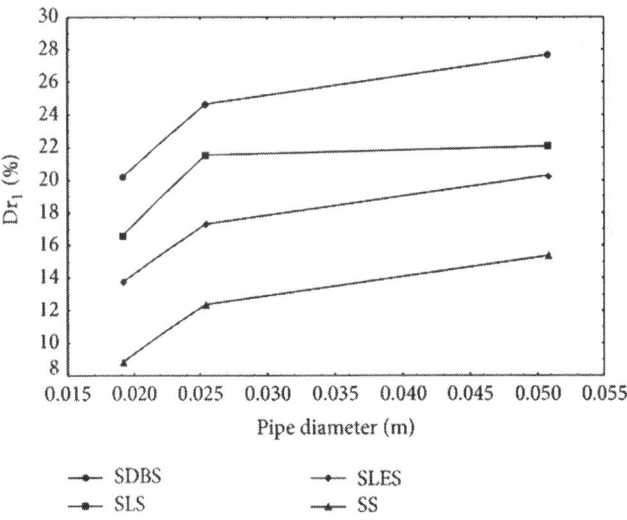

Figure 3: Effect of pipe diameter on drag reduction for 50 ppm surfactant concentration and 6 m³/h flow rate.

Effect of Flow Rate

Figure 4 shows the effect of solution velocity (v) on the percentage drag reduction (%Dr) in terms of dimensionless group (Re). The results show that the drag reduction percentage increases with increasing fluid velocity. Increasing the fluid velocity means increasing the degree of turbulence inside the pipe, this will provide a better media to the drag reducer to be more effective. The behavior of increasing %Dr with velocity of fluid may be explained due to relation between degree of turbulence controlled by the solution velocity and the additive effectiveness. The same results were obtained by Kim et al. [12]; the drag reduction was larger at high Reynolds number.

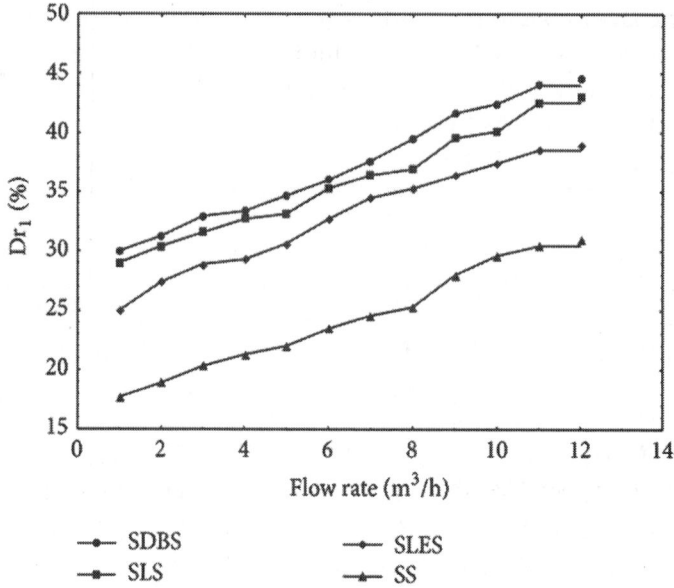

Figure 4: Effect of flow rate on drag reduction for different surfactants at 250 ppm concentration flowing in 0.0254 m I.D. pipe.

Effect of Friction

Figure 5 showed that the friction factor for various Re, pipe diameter, additives type, and additives concentrations is shown. These figures are divided into four regions. These regions are as follows [13].

- Laminar flow region (Re < 2300), where the friction factor follows Poisuell's law as follow:

$$f = 16\text{Re}^{-1}.$$

(3)

- Transition region (Re = 2300–3000), where the flow change from laminar to turbulent flow. Friction coefficient rises rapidly.
- Turbulent region (Re > 3000), where the friction factor follow Blasius law:

$$f = 0.0791\text{Re}^{-0.25}.$$

(4)

- Virk asymptote region, which is suggested by Virk to represent the greatest possible fall in resistance in which the relation between friction factor (f) and Re does not depend on the nature of the additives or pipe diameter. The formula for Virk is

$$f = 0.59 Re^{-0.58}.$$

(5)

These figures showed that the friction factor decreased with decreasing the pipe diameter, with increasing concentration of additives, and with increasing fluid velocity. From these figures, it can be noticed that most of experimental data points are located at or close to Blasius asymptote when the solvent was pure. After the addition of additives, the data points positioned toward Virk asymptote which represent the maximum limits of drag reduction (Table 4). It was difficult to reach these limits of lowering resistance because the higher concentration of additives are required to achieve this condition. But it must be taken into account that higher concentration should not affect solvent properties.

Table 4: Maximum values of %Dr and %Fl at 250 ppm concentration surfactant

Additive type	Pipe diameter (m)	Flow rate (m³/hr)	Max. %Dr$_1$	Max. %Dr$_2$	Max. %Fl$_1$	Max. %Fl$_2$
SDBS	0.0508	12.00	48.29	54.48	43.73	54.17
SLS	0.0508	12.00	44.46	47.30	38.19	42.23
SLES	0.0508	12.00	41.39	43.00	34.16	36.23
SS	0.0508	12.00	33.36	35.34	25.01	27.10
SDBS	0.0254	12.00	45.31	42.54	39.36	35.63
SLS	0.0254	12.00	42.48	40.51	35.55	33.06
SLES	0.0254	12.00	38.40	36.98	30.54	28.91
SS	0.0254	12.00	32.00	28.80	23.63	20.54
SDBS	0.0191	6.00	31.50	29.86	23.13	21.54
SLS	0.0191	6.00	27.40	26.58	19.26	18.52
SLES	0.0191	6.00	25.70	24.87	17.75	17.03
SS	0.0191	6.00	21.37	20.00	14.14	13.06

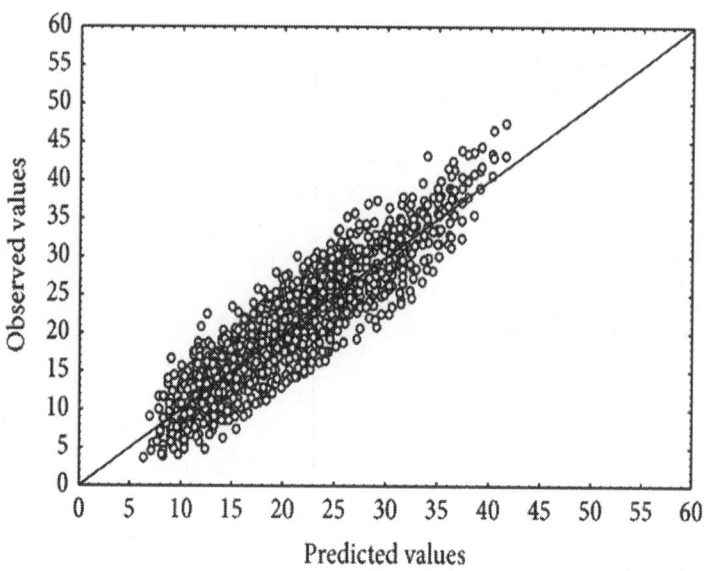

Figure 5: Predicted versus observed values of drag reduction for crude oil flowing through pipes.

Correlation of Variables

The dimensional analysis was used in the present work for grouping the significant quantities into a dimensionless group to reduce the number of variables appearing and to make the result so compact and applicable to all similar situations. The drag reduction is influenced by the physical properties of solvent and properties of flow. The relationship may be written as

$$\Delta P = f\left(D, \mu, \rho, V, C, L, \varepsilon\right).$$

(6)

By applying the dimensional analysis, the following nondimensional relation was proposed:

$$\%\mathrm{Dr} = f\left(\mathrm{Re}, \frac{\varepsilon}{d}, \frac{L}{d}, C\right)$$

(7)

or

$$\%\mathrm{Dr} = a(\mathrm{Re})^{b}\left(\frac{\varepsilon}{d}\right)^{c}\left(\frac{L}{d}\right)^{d}(\mathrm{C})^{k}. \tag{8}$$

The method of least square was used to determine the coefficients of correlation for Reynolds number range (4341–72775). The coefficients for this system (i.e., for pipes diameter, additives type, and solvents type) were summarized in the following with 0.9032 correlation coefficient:

$$\%\mathrm{Dr} = 0.134(\mathrm{Re})^{0.324}\left(\frac{\varepsilon}{d}\right)^{-0.806}\left(\frac{L}{d}\right)^{-0.934}(\mathrm{C})^{0.153}. \tag{9}$$

CONCLUSIONS

The additives (SDBS, SLS, SLES, and SS) were found to be effective drag reducing agent when used with Kirkuk crude oil. Drag reduction percent or flow increase percent are increased as the velocity of solution increased. Drag reduction percent is increased with increasing concentration of additives. It is observed that the additives do not affect the physical properties of used crude oils. A correlation equation was obtained to represent the experimental data mathematically using least square method in analysis. This correlation showed the drag reduction percent (%Dr) as a function of Reynolds number (Re), concentration of additives (C), roughness factor (ε/d), and the ratio (L/d). The results showed good agreement between the observed drag reduction percent values and the predicted ones with high value of correlation coefficients.

ACKNOWLEDGMENTS

This work was supported by Baghdad University, Chemical Engineering Department, which is gratefully acknowledged.

REFERENCES

1. F.-C. Li, Y. Kawaguchi, B. Yu, J.-J. Wei, and K. Hishida, "Experimental study of drag-reduction mechanism for a dilute surfactant solution flow," International Journal of Heat and Mass Transfer, vol. 51, no. 3-4, pp. 835–843, 2008.

2.   K. Prajapati, Interactions between drag reducing polymers and surfactants [M.S. thesis], University of Waterloo, Ontario, Canada, 2009.

3.   H. Ferhat and G. Sylvain, "Drag reduction by surfactant in closed turbulent flow," International Journal of Engineering Science and Technology, vol. 2, pp. 6876–6879, 2010.

4.   C. B. Lester, "Drag reduction agents," Oil and Gas Journal, vol. 4, pp. 51–56, 1985.

5.   R. Martínez-Palou, M. D. L. Mosqueira, B. Zapata-Rendón et al., "Transportation of heavy and extra-heavy crude oil by pipeline: a review," Journal of Petroleum Science and Engineering, vol. 75, no. 3-4, pp. 274–282, 2011. · ·

6.   J. Zakin, "Surfactant drag reduction," Reviews in Chemical Engineering, vol. 1, pp. 252–320, 1998.

7.   R. Darby, Engineering Fluid Mechanics, Marcel Dekker, New York, NY, USA, 2nd edition, 2001.

8.   F. A. Holland and R. Bragg, Fluid Flow for Chemical Engineers, Edward Arnold, London, UK, 2nd edition, 1995.

9.   S. Takashi and U. Hiromoto, "Drag reduction and heat transfer reduction by cationic surfactants," Journal of Chemical Engineering of Japan, vol. 26, no. 1, pp. 103–106, 1993. ·

10.   D. Mowla and A. Naderi, "Experimental study of drag reduction by a polymeric additive in slug two-phase flow of crude oil and air in horizontal pipes," Chemical Engineering Science, vol. 61, no. 5, pp. 1549–1554, 2006. · ·

11.   H. R. Karami and D. Mowla, "Investigation of the effects of various parameters on pressure drop reduction in crude oil pipelines by drag reducing agents," Journal of Non-Newtonian Fluid Mechanics, vol. 177-178, pp. 37–45, 2012.

12.   N.-J. Kim, J.-Y. Lee, S.-M. Yoon, C.-B. Kim, and B.-K. Hur, "Drag reduction rates and degradation effects in synthetic polymer solution with surfactant additives," Journal of Industrial and Engineering Chemistry, vol. 6, no. 6, pp. 412–418, 2000.

13.   S. N. Ashrafizadeh, E. Motaee, and V. Hoshyargar, "Emulsification of heavy crude oil in water by natural surfactants," Journal of Petroleum Science and Engineering, vol. 86, pp. 137–143, 2012.

Chapter 8

Simulation and Experimental Investigation of Thermal Performance of a Miniature Flat Plate Heat Pipe

R. Boukhanouf[1] and A. Haddad[2]

[1]Department of the Built Environment, University of Nottingham, Nottingham NG7 2RD, UK

[2]FrigoDynamics GmbH, Bahnhofstraße 16, 85570 Markt Schwaben, Germany

ABSTRACT

This paper presents the results of a CFD analysis and experimental tests of two identical miniature flat plate heat pipes (FPHP) using sintered and screen mesh wicks and a comparative analysis and measurement of two solid copper base plates 1mm and 3mm thick. It was shown that the design of the miniature FPHP with sintered wick would achieve the specific temperature gradients threshold for heat dissipation rates of up to 80W. The experimental results also revealed that for localized heat sources of up to 40W, a solid copper base plate 3mm thick would have comparable heat transfer performances to that of the sintered wick FPHP. In addition, a marginal effect on the thermal performance of the sintered wick FPHP was recorded when its orientation was held at 0°, 90°, and 180° and for heat dissipation rates ranging from 0 to 100W.

INTRODUCTION

Conventional heat sink-fan air coolers in electronics packages are becoming inadequate for use in faster, compact, and more powerful multitasked microprocessors that generate large quantities of attendant heat. This has led current research to focus on high-performance and compact thermal solutions [1]. Heat pipes have been extensively researched and applied in various embodiments for electronics cooling ranging from simple cylindrical geometries to complex configurations [2]. FPHPs in particular have high-heat transfer capability, can maintain a uniform temperature over the evaporator surface when densely packed with heat-generating components, and decrease the thickness of finned heat sinks base material [3–6].

Recent research by Christensen and Graham [7] investigated the performance of heat sinks in packaging high-power (>1W) light-emitting diode (LED) arrays and concluded that flat plate heat pipes form an important thermal component to achieve long operating life and high reliability. Huang and Liu [8] demonstrated analytically the increased capability of mounting a localized heat source and heat sink on the same surface of an FPHP. Similarly, Qin and Liu [9] investigated liquid flow in an anisotropic permeability wick of a flat plate heat pipe, determining the effect of heat source location on fluid distribution in the inside of the heat pipe. Further work on finding the optimum location of mounting multiple heat sources on an FPHP evaporator surface was demonstrated by Tan et al. [10] through a simplified analytical solution to a two-dimensional pressure and velocity distribution within the wick. Recently, Sonan et al. [11] developed a simulation model for the transient thermal performance of a 40 × 40 × 0.9mm FPHP with specific applications to cooling multiple electronics components where space restriction imposes that heat sources and heat sinks need to be mounted on the same surface. The above research agrees that a cost premium associated with a well-designed FPHP for advanced thermal management solution in electronics cooling should be reflected in its superior thermal performance compared to solid copper or aluminium base materials of similar dimensions.

This paper investigates the design and thermal performance of a miniature FPHP as an effective supporting shelf for printed circuit boards (PCB) of radio frequency (RF) components to dissipate and transport heat away to the aluminium enclosure. In this paper, two identical miniature FPHPs configurations one with a sintered copper powder and the other one with a screen copper mesh wick were designed and tested. A further benchmarking exercise, using both CFD analysis and experimental measurements, was carried out by comparing the thermal performance of the FPHP to that of a monolithic solid copper plate with similar dimensions.

DESCRIPTION OF THE ELECTRONICS ENCLOSURE COOLING SYSTEM

The work addresses the cooling requirement to maintain a specified temperature limit for heat-generating RF components that are housed in an existing aluminium enclosure as part of a large telecommunication control system. The aluminium enclosure consists of two separate copper shelves and an air cooled finned base plate, as shown in Figure 1. The PCB of high heat dissipating RF components was mounted on the lower shelf of the enclosure to allow for direct contact with the base heat metal spreader while low-power rated RF components were placed on the top shelf. In the original design, the enclosure's shelves were made of 1mm thick copper base. However, frequent and premature failures of RF components prompted the review of the enclosure's thermal performance. Hence, a redesign of the copper shelves to keep the operating temperature gradients of the RF components within the specified limits was performed using both CFD simulation and an experimental validation analysis. The work consists in investigating the thermal performance of the enclosure's shelves that are made of a 1mm and 3mm solid copper base plates and two miniature FPHPs of 5mm overall height with one using sintered copper powder wick and the other one a screen mesh wick.

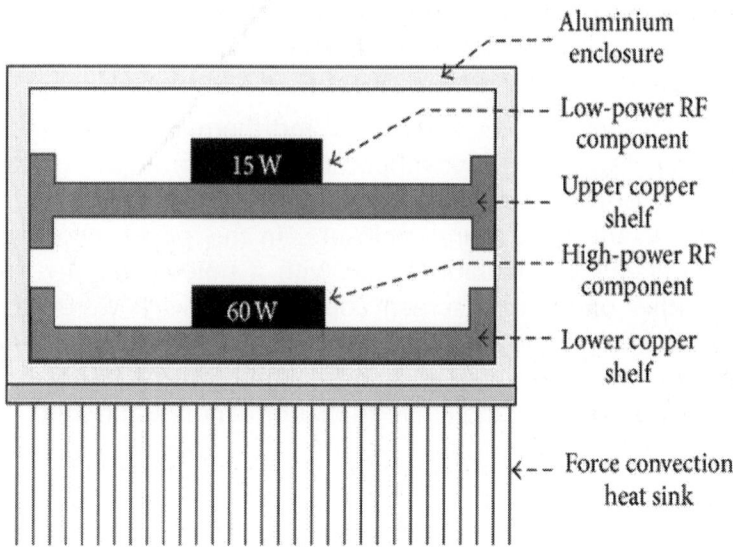

Figure 1: Schematic of the electronics cooling enclosure.

FPHP Design

The FPHP base material was made of copper material and water was selected as the working fluid for its compatibility and suitable operating temperature range. The outer shell of the constructed prototype miniature FPHPs for the sintered and mesh wicks is shown in Figure2 (a). The inner structure of the sintered wick FPHP with the condenser cover plate removed to reveal the sintered wick layer and erected pillars on the evaporator surface is shown in Figure2(b). Similarly, the structure of the miniature FPHP with copper mesh wick is shown in Figure2(c). The design of the mesh wick FPHP was adopted from Bakke [12] and Rosenfeld et al. [13], which consists of using a fine mesh layer to provide capillary pumping force of the working fluid and a coarse mesh to support and maintain the structural integrity of the vapour space.

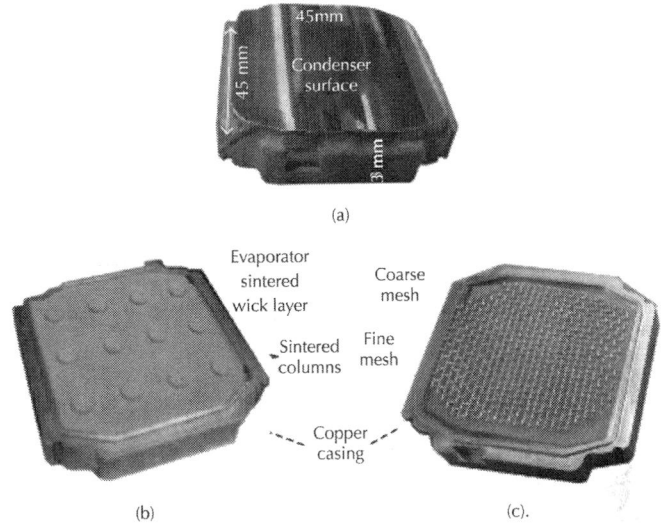

Figure 2: Constructed miniature FPHP (a) sealed unit, (b) sintered wick, and (c) mesh wick.

FPHP CFD Simulation

The thermal performance of the miniature FPHP was simulated using FloTHERM, a commercial CFD simulation software. The computer simulation includes analysing the complex flow pattern of the working fluid in the wick and establishing the temperature profiles in the FPHP evaporator. Flotherm is a finite volume-based software package that uses simple Cartesian grid meshing and has built in boundary conditions for common heat transfer devices. The rectangular shape of the FPHP, heater block, and cold plate lend themselves well to meshing using Cartesian coordinates and hence the use of Flotherm for a fast and converging solution. Description of the mathematical model for momentum, mass, and energy conservation that underpin the CFD simulation was not the focus of this paper as similar models are widely available in published literature that can be found in [14–16].

The Flotherm model was built using standard Cuboids and Prism elements for the FPHP components including the evaporator and condenser copper plates, the wick layer, the void (vapour) space, and the supporting solid columns. Planar resistance object model was used to define the thermal properties for each object. This is a useful tool where the thermal resistance of an object can be inserted manually or determined from other thermal parameters such as the thermal conductivity and heat transfer coefficient. Modelling of the porous wick layer, in particular, requires prior knowledge of the permeability of the porous wick structure. The volumetric flow rate, V, of the working fluid was also required as an input parameter in the simulation. This was calculated from the following relationship:

$$\dot{V} = \frac{Q_H}{\rho_l h},$$

(1)

Where Q_H is the rate of heat generation in the heat source and h, and ρ_l are the latent heat and density of the working fluid, respectively. In addition, it is well known that the failure of heat pipes is often attributed to operation beyond the device's wick capillary limit. This can be obtained by characterizing the actual pressure of the liquid in the wick pores under different heat flux levels using Laplace-Young equation as follows [1]:

$$\Delta p_c = \frac{2\sigma_l}{r_{pore}},$$

(2)

Where p_c is the capillary pressure drop in the wick, ρ_l is the surface tension of the liquid, and r_{pore} is the pore radius of the wick.

Equally, the effective thermal conductivity of a saturated wick, k_{eff}, was obtained from the following expression [17, 18]:

$$k_{eff} = \beta \left(\varphi k_l + (1 - \varphi) k_s \right)$$

$$+ \frac{(1 - \beta)}{\varphi / k_l + (1 - \varphi) / k_s},$$

(3)

Where φ is the porosity of the wick, k_l and k_s are the thermal conductivity of liquid water and solid wick, respectively. According to Bhattacharya et al. [17], the best-fit data for measuring the effective thermal conductivity of a porous material is for $\beta = 0.35$ with an overall R^2 value of 0.97. This is consistent with measured thermal conductivities of about 40 W/(mK) in heat pipes with fully saturated wicks, while the vapour space is considered to have a very large heat transfer coefficient of the order of 50000W/m²K [16, 18].

A schematic representation of the sintered wick heat pipe given in Figure 3 illustrates the location of the heat source, a cross section of the wick layer, the working fluid circulation paths, and the condenser cold plate heat sink.

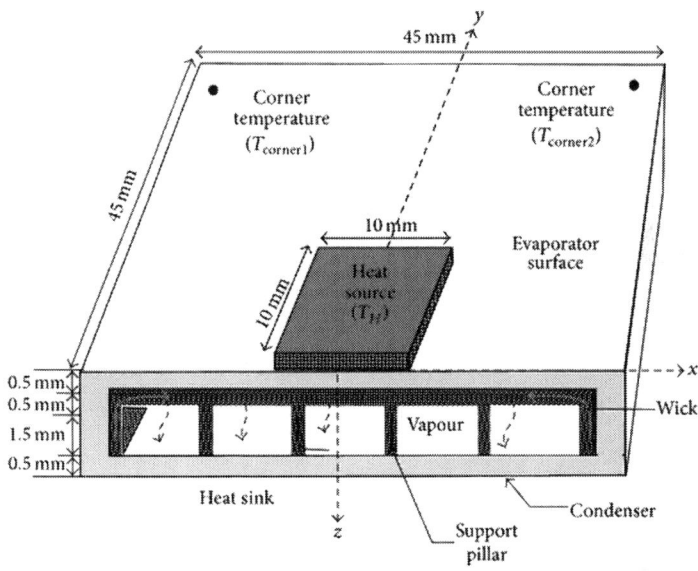

Figure 3: A schematic representation of the FPHP.

In this analysis, it was also assumed that the heat source is of constant heat flux type which is applied to the evaporator base plate (at z=0) immediately under the heat source while the remaining outer surface of the evaporator including the edge walls is considered to be adiabatic. In the inner section of the heat pipe, it was assumed that the liquid-vapour interface temperature is equal to the vapour saturation temperature of the working fluid that can be calculated from Clausius-Clapeyron relationship [19]. Similarly, the liquid velocity at the interface of the wick layer-evaporator wall was assumed to equal zero. At the condenser-vapour interface, the temperature of the condenser section was maintained at 35°C using a chilled liquid cold plate with a heat transfer coefficient in the order of 2000W/ (m²K).

A thermocouple placed at the interface surface between the evaporator and heat source was used to measure the temperature of the heater block, T_H, and a further two thermocouples were placed at the two farthest corners on the evaporator to measure $T_{corner1}$ and $T_{corner2}$. Furthermore, the heat dissipation from the RF components was simulated using an electric heater cartridge that is inserted in a solid aluminium block of 10mm × 10mm with a controlled heat dissipation rates. The condenser surface was maintained to the desired temperature of 35°C by clamping directly onto its surface a chilled water cold plate. The main design properties of the miniature FPHPs with sintered and mesh wicks are given in Table 1.

Table 1: Design properties of the sintered and mesh wick miniature FPHP

Casing	
Material	Copper
Height	3mm
Dimension	45mm × 45mm
Heater block size	10mm × 10mm
Working fluid	Water
Sintered copper powder wick	
Wick thickness	0.5mm
Vapour space height	1.5mm
Porosity	50%
Pore radius	40µm
Permeability	$1.43 \times 10^{-11} m^2$
Screen copper mesh wick	
Fine mesh material	Phosphor bronze 320 mesh/in
Wire diameter	0.03mm
Porosity	42%
Wick thickness	1mm
Supporting coarse mesh material	Phosphor bronze 16 mesh/in

CFD Simulation Results

The CFD simulation was used to evaluate the sintered wick FPHP thermal performance by analysing the steady state liquid flow pressure and velocity distribution in the sintered wick as shown in Figure 4. The speed of the liquid flow in the wick structure is presented by arrows pointing towards the heat source (dashed line square) and arranged by color in contours of equal speeds. It can be seen that the speed of the liquid flow is lowest at regions most distant from the heat source (contours of purple arrows) and increases gradually as the liquid is drawn towards the heat source (contours of orange and red arrows) to replenish the evaporated liquid from the wick pores in the constant heat flux section. The liquid flow speed then drops sharply in the region immediately underneath the heat source as the liquid evaporates from the wick. The effect of the sintered pillars is also visible in that the liquid flow speed field contours are altered in a way that high-fluid speed spots were developed around the pillars. Furthermore, Figure 4 shows the liquid pressure distribution in the wick with the high pressure region (in red color) away from the heat source and the low-pressure regions (in blue/purple color) immediately beneath the heat source.

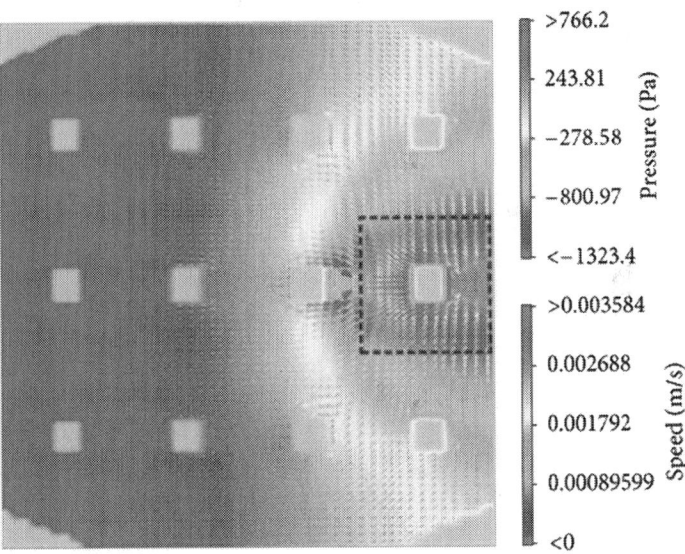

Figure 4: CFD simulation of speed and pressure fields of the liquid in the FPHP-sintered wick (Q_H=80W).

From the CFD analysis, it was found that the total pressure difference generated by the capillary forces of the wick is 2089.6N/m². This is markedly lower than the capillarity pumping limit of the wick of 3952N/m² which was calculated using (2) under the working conditions given in Table 2.

Table 2: Saturated wick operating properties

Heat load	Liquid surface tension	Liquid density	Liquid latent heat of vaporisation	Sintered powder wick permeability	Liquid flow rate from (1)	Pressure drop from (2)	Pressure drop (CFD)
Q_{H} (W)	σ_{l} (kg/s²)	ρ_{L} (kg/m³)	h (kJ/kg)	ε_{wick} (m²)	V (m³/s)	P_{c} (N/m²)	P_{CFD} (N/m²)
80	0.0626	972	2310	1.43E-11	3.563E-08	3952	2089.6

EXPERIMENTAL SETUP AND RESULTS

The experimental rig setup to test the thermal performance of the miniature FPHP and solid copper base samples is shown in Figure 5. The rig was equipped with a liquid filling and venting system for charging the FPHP, a rotating beam for mounting the heat pipe at different tilt angles (0° to 180°), a variable power supply to control heat dissipation from the heat source, a chilled water supply to control the condenser temperature, and associated sensors and data acquisition equipment. The miniature FPHP test sample was clamped onto the tilting beam with 0° angle being designated for operation against gravity (i.e., evaporator is above the condenser). An electric heater cartridge placed inside an aluminium block of 10mm × 10mm was used as heat source which heat dissipation rate was controlled by a variable power supply. A thermal interface material 0.5mm thick was used as an interface between the FPHP test sample, the heater block, and the condenser cold plate to minimize the contact thermal resistances.

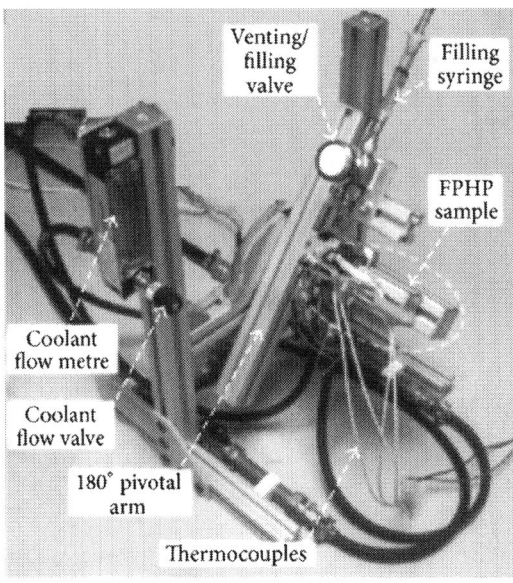

Figure 5: The experimental test rig setup.

The performance of the FPHP test sample was evaluated at various heat dissipation rates by increasing the heat source power at equal increments of 10W to a maximum of 100W or until the temperature of the evaporator surface reached a threshold of 100°C. At each heat input increment the temperature of the evaporator surface was allowed to reach steady state. In addition, the effect of orientation on the FPHPs was evaluated by repeating the experimental measurements at tilt angles of 0°, 90°, and 180°. For comparison, further experimental tests were conducted on a 1mm and 3mm thick solid copper base plates under identical controlled conditions. In all tested samples, the operating temperature was measured at the heat source, T_H, evaporator plate corners $T_{corner1}$ and $T_{corner2}$, and at the condenser surface, T_{cond}.

Solid Copper Base Plate Shelves

The initial tests of two solid copper plates of 45mm × 45mm and a thickness of 1mm and 3mm were carried out to provide a benchmark data which the thermal performance of the FPHPs was compared to. Results of these tests are shown in Figure 6, where it can be seen that for the same heat dissipation rates the temperature of the heat source in the 1mm thick copper base is higher than that of the 3mm thick copper base, particularly at high heat dissipation rates. For example, at a heat dissipation rate of 80W, the recorded temperature, T_H, is 84°C and 73°C for the 1mm and 3mm thick base plates, respectively. Similarly, the temperature gradients between the heat source and the two far end corners, $T_{H-corner}$, and between the heat source and the heat sink, T_{cond}, are approximately 15°C higher for the 1mm thick copper base than for its 3mm thick counterpart. This shows that the 3mm thick copper base plate has superior heat-spreading properties as predicated in the CFD simulation results.

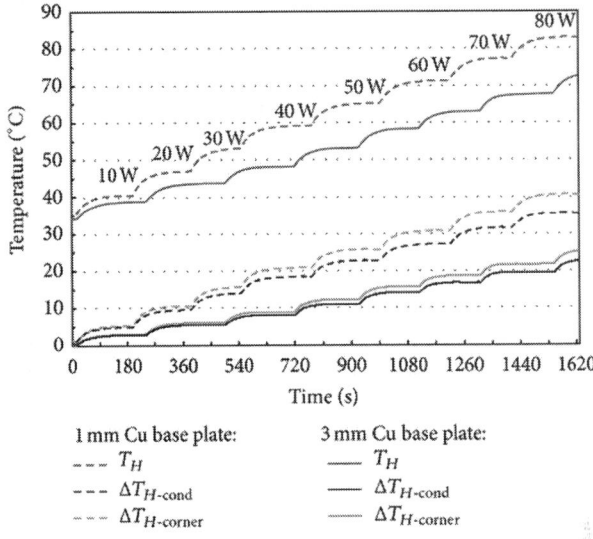

Figure 6: Temperature variation of a 1mm and 3mm thick copper base plates.

Screen Mesh Wick Miniature FPHP

The procedure for evaluating the thermal performance of the screen mesh wick FPHP was similar to that presented in previous case with additional experimental measurements to assess the effect of orientation at tilt angles of 0°, 90°, and 180°. The measured temperature changes at the heat source and across the evaporator surface are shown in Figures 7(a) and 7(b). These show that at 0° and 90° tilt angles the FPHP performance is only comparable to that of the 1mm thick solid copper base. For example at high-heat dissipation rates (80W), the temperature of the heat source approaches 90°C and a large temperature gradient (30 to 40°C) appears across the evaporator surface and between the heat source and heat sink, $T_{H\text{-cond}}$. The thermal performance of the heat pipe has improved marginally for a tilt angle of 180° (gravity assisted wick capillary forces), but it remains that the 3mm solid copper base plate performed better. This unexpected poor thermal performance may be attributed to the process of fabrication in which poor contact between the fine screen mesh and the inner evaporator wall could have prevented liquid circulation, leading to a high interfacial thermal resistance.

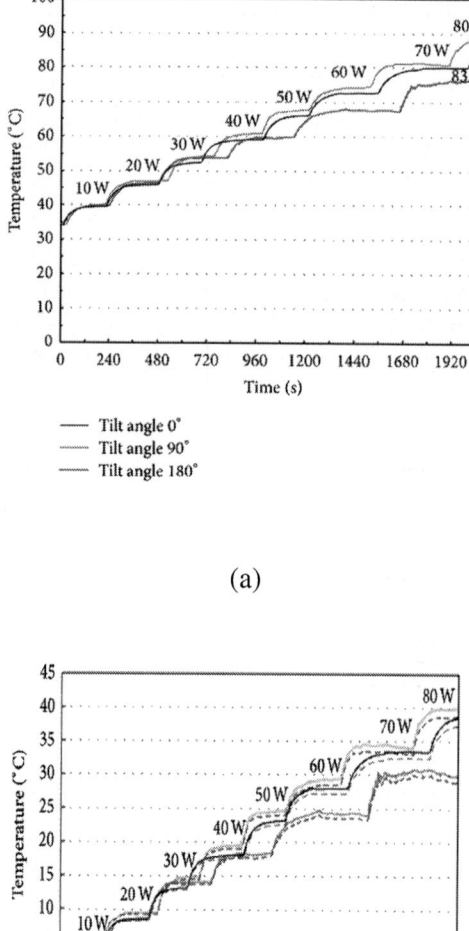

Figure 7: Screen mesh wick FPHP temperature variation at various heat dissipation levels and tilt angles: (a) heat source temperature, (b) evaporator surface temperature gradient.

Sintered Copper Powder Wick FPHP

The sintered wick FPHP thermal performance is shown in Figures 8(a) and 8(b). It can be seen that for a heat dissipation rate of 80W the measured temperature of the heat source at a tilt angle of 0° is 69°C, which is lower by 15°C and 5°C compared to the 1mm and 3mm copper base plates, respectively. For a tilt angle of 180° the temperature of the heat source has dropped even further to 61°C, an improvement of 12°C compared to the 3mm copper base plate. Similarly, the measured temperature gradients $T_{H\text{-}cond}$ and $T_{H\text{-}corner}$ are 6°C and 5°C lower than that of the 3mm copper base plate for a tilt angle of 0° and 10°C and 8°C for a tilt angle of 180°, respectively.

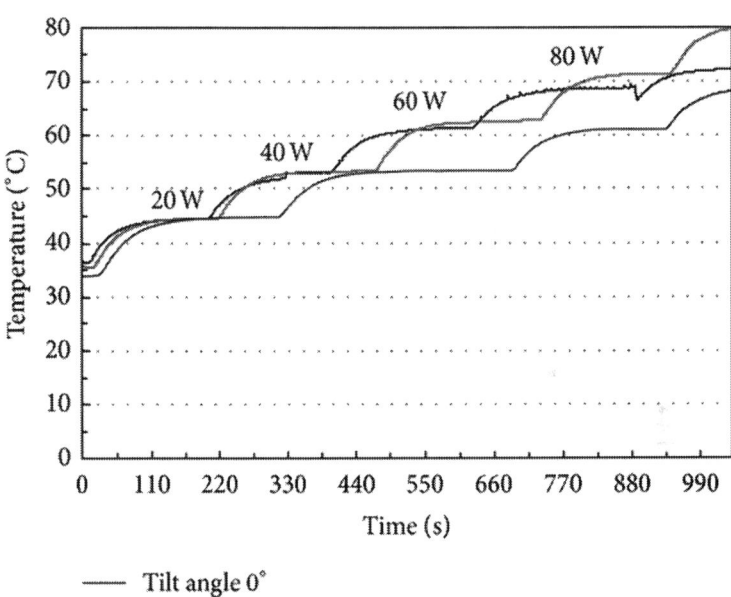

(a)

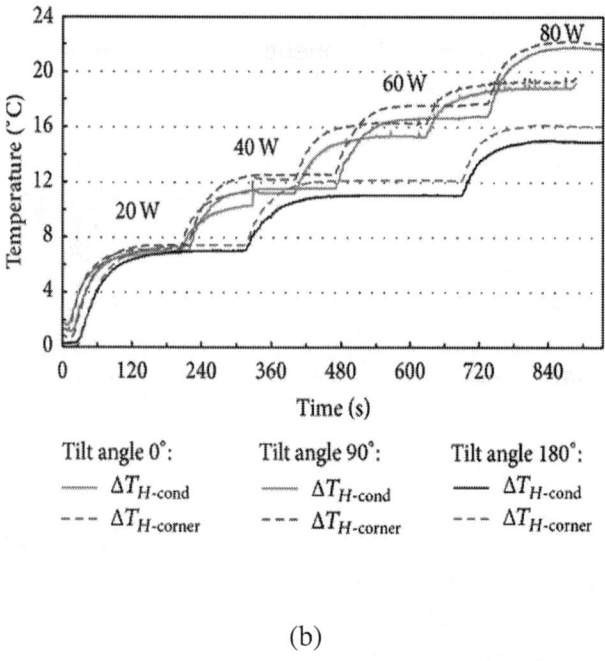

Tilt angle 0°:   Tilt angle 90°:   Tilt angle 180°:

— $\Delta T_{H\text{-cond}}$ — $\Delta T_{H\text{-cond}}$ — $\Delta T_{H\text{-cond}}$

--- $\Delta T_{H\text{-corner}}$ --- $\Delta T_{H\text{-corner}}$ --- $\Delta T_{H\text{-corner}}$

(b)

Figure 8: Sintered wick FPHP temperature at various heat dissipation levels and tilt angles: (a) heat source temperature, (b) surface temperature gradient.

The high-thermal performance of the sintered wick miniature is due to high heat conduction and spreading capability of the sintered wick compared to a simple monolithic solid copper base plate or screen mesh wick FPHP.

EVALUATION OF BULK AND THERMAL SPREADING RESISTANCE

The effective thermal resistance of an electronics component mounted on a PCB is the sum of a one-dimensional bulk resistance and a thermal spreading resistance. The one-dimensional bulk thermal resistance is expressed as follows [4, 20, and 21]:

$$R_b = \frac{T_H - T_{\text{cond}}}{\dot{Q}_H}.$$

(4)

The thermal spreading resistance is associated, however, with discrete heat-generating components when mounted on a cold base plate, as found in electronics packages. The thermal spreading resistance characterizes the ability of a base plate to spread the heat uniformly across the base plate surface (or the evaporator surface in the case of FPHP). The thermal spreading resistance could be of a similar magnitude to the one-dimensional bulk resistance in some designs of heat exchangers. Hence, its omission can lead to significant errors in estimating the temperature of a PCB, resulting in components overheating and failing prematurely. The thermal spreading resistance, R_{sp}, is computed using the following expression [21]:

$$R_{sp} = \frac{T_H - \left(T_{corner1} + T_{corner2}\right)/2}{\dot{Q}_H}. \tag{5}$$

The variation of the one-dimensional bulk and thermal spreading resistances for the tested copper base plates and miniature FPHPs are shown in Figure 9. It can be seen that for heat dissipation rates of up to 40W, the sintered wick heat pipe and the 3mm thick copper base plate have comparable thermal performances in that both the one-dimensional and spreading thermal resistances are of similar magnitude. For higher heat dissipation rates, however, the advantage of a sintered wick FPHP becomes more apparent as the one-dimensional bulk and thermal spreading resistances are lower compared to other designs.

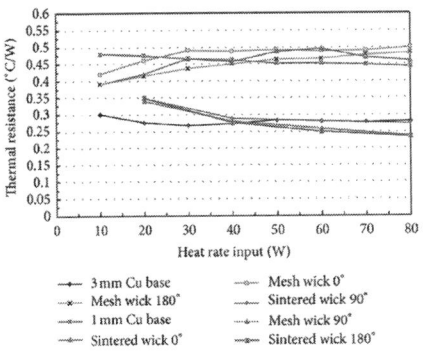

(a)

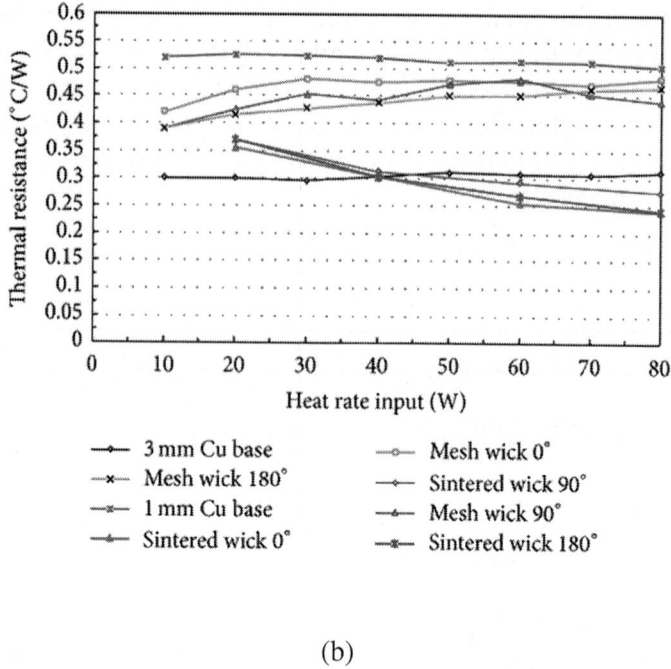

(b)

Figure 9: Thermal resistance: (a) one dimensional resistance, (b) spreading resistance.

Finally, the uncertainty error of calculating the one-dimensional bulk and thermal spreading resistances is estimated from the measured data and the accuracy of the instruments used in the experiments. The accuracy of the T-type thermocouples is 0.5°C while the average error for power supply reading (wattmeter) is estimated at 2.3W. Therefore, using the single sample analysis [22], the relative uncertainty error of the thermal resistance is calculated as follows:

$$e_R = \sqrt{\left(\frac{e_T}{\Delta T}\right)^2 + \left(\frac{e_Q}{\dot{Q}_H}\right)^2}.$$

(6)

It was assumed that the electrical power input to the heater block was fully dissipated as heat energy and that heat loss by convection and radiation from the test sample was negligible. The relative uncertainty error calculation of the thermal resistances has been limited to the case of the sintered wick FPHP, as shown in Table 3. The uncertainty calculation of the FPHP thermal resistance decreases from 13.6% to 3.8% for heat input rates of 20W and 80W, respectively. Although the average reading accuracy of the wattmeter is 2.3W, the reading scale is nonlinear and the error of measurement is highest at low end of the scale, leading to large uncertainty of the measured thermal resistance for a heat rate input of 20W.

Table 3: Sintered wick FPHP thermal resistance calculation uncertainty

Heat input (W)	20	40	60	80
Calculated thermal resistance uncertainty (%)	13.60	6.00	4.14	3.80

CONCLUSIONS

This work investigated the thermal performance of a miniature FPHPs with sintered and screen mesh wicks for application in electronics cooling. The thermal performance of the FPHPs was further compared to that of copper solid base plates 1mm and 3mm thick. The ability of each sample to dissipate heat was evaluated by measuring the temperature distribution on the mounting surface and the temperature gradient between the heat source and heat sink. The main findings can be summarized as follows.

- The CFD results of predicting that the sintered wick FPHP would perform better than other arrangements were in good agreement with the experimental measurements.

- It was found that the 3mm thick copper base plate thermal performance surpasses that of the 1mm and achieves higher heat conduction and spreading performance than the screen mesh wick FPHP.

- The 3mm thick copper base can perform adequately with heat dissipation rates of up to 40W.

- For heat dissipation rates higher than 40W, the sintered wick FPHP outperforms the 3mm copper base plate and its application would justify its high cost.

- The temperature measured on the evaporator surface of the sintered wick FPHP shows that there is no sign of liquid dry-out conditions in the wick for the range of heat dissipation rates.
- The orientation of the sintered wick FPHP had marginal effect on its performance.

ACKNOWLEDGMENTS

The authors wish to thank EPSRC (Engineering and Physical Sciences Research Council) for its financial support of the project under Grant EP/P500389/1 and Thermacore Europe Ltd for providing financial and technical help.

REFERENCES

1.   D. A. Reay and P. A. Kew, Heat Pipes: Theory, Design and Applications, Butterworth-Heinemann, New York, NY, USA, 5th edition, 2006.

2.   M. Groll, "Heat pipe research and development in western Europe," Heat Recovery Systems and CHP, vol. 9, no. 1, pp. 19–66, 1989.

3.   A. Basiulis, H. Tanzer, and S. McCabe, "Thermal management of high power PWB's through the use of heat pipe substrates," in Proceedings of the 6th Annual International Electronics Packaging Conference, vol. 6, p. 501, San Diego, Calif, USA, 1986.

4.   M. Adami and B. Yimer, "Development and evaluation of a planar heat pipe for cooling electronic systems," Chemical Engineering Communications, vol. 90, no. 1, pp. 57–74, 1990.   ·

5.   S. W. Kang, S. H. Tsai, and H. C. Chen, "Fabrication and test of radial grooved micro heat pipes," Applied Thermal Engineering, vol. 22, no. 14, pp. 1559–1568, 2002.   ·

6.   C. Y. Liu, C. Y. Liu, K. C. Leong, Y. W. Wong, and F. L. Tan, "Performance study of flat plate heat pipe," in Proceedings of the International Conference on Energy and Environment (ICEE), pp. 512–518, Begell House Inc., New York, NY, USA, 1996.

7. A. Christensen and S. Graham, "Thermal effects in packaging high power light emitting diode arrays," Applied Thermal Engineering, vol. 29, no. 2-3, pp. 364–371, 2009. ·

8. X. Y. Huang and C. Y. Liu, "The pressure and velocity fields in the wick structure of a localized heated flat plate heat pipe," International Journal of Heat and Mass Transfer, vol. 39, no. 6, pp. 1325–1330, 1996. · ·

9. W. Qin and C. Y. Liu, "Liquid flow in the anisotropic wick structure of a flat plate heat pipe under block-heating condition," Applied Thermal Engineering, vol. 17, no. 4, pp. 339–349, 1997.

10. B. K. Tan, X. Y. Huang, T. N. Wong, and K. T. Ooi, "A study of multiple heat sources on a flat plate heat pipe using a point source approach," International Journal of Heat and Mass Transfer, vol. 43, no. 20, pp. 3755–3764, 2000. · ·

11. R. Sonan, S. Harmand, J. Pellé, D. Leger, and M. Fakès, "Transient thermal and hydrodynamic model of flat heat pipe for the cooling of electronics components,"International Journal of Heat and Mass Transfer, vol. 51, no. 25-26, pp. 6006–6017, 2008. ·

12. A. P. Bakke, "Light weight rigid flat plate heat pipe utilizing copper foil container laminated to heat treated Aluminium plates for structural stability," US Patent No. 6679318 B2, 2004.

13. J. H. Rosenfeld, N. J. Gernert, D. V. Sarraf, P. Wollen, F. Surina, and J. Fale, "Flexible heat pipe," US Patent No. 6446706 B1, 2002.

14. Y. Koito, H. Imura, M. Mochizuki, Y. Saito, and S. Torii, "Numerical analysis and experimental verification on thermal fluid phenomena in a vapor chamber," Applied Thermal Engineering, vol. 26, no. 14-15, pp. 1669–1676, 2006. · ·

15. G. Carbajal, C. B. Sobhan, G. P. Bud Peterson, D. T. Queheillalt, and H. N. G. Wadley, "A quasi-3D analysis of the thermal performance of a flat heat pipe," International Journal of Heat and Mass Transfer, vol. 50, no. 21-22, pp. 4286–4296, 2007. · ·

16. R. Ranjan, J. Y. Murthy, and S. V. Garimella, "Analysis of the wicking and thin-film evaporation characteristics of microstructures," Journal of Heat Transfer, vol. 131, no. 10, pp. 1–11, 2009. · ·

17. A. Bhattacharya, V. V. Calmidi, and R. L. Mahajan, "Thermophysical properties of high porosity metal foams," International Journal of Heat and Mass Transfer, vol. 45, no. 5, pp. 1017–1031, 2002. ·
·

18.   J. Thayer, "Analysis of a heat pipe assisted heat sink," in Proceedings of the 9th International FLOTHERM Users Conference, Orlando, Fla, USA, October 2000.

19.   B. Xiao and A. Faghri, "A three-dimensional thermal-fluid analysis of flat heat pipes,"International Journal of Heat and Mass Transfer, vol. 51, no. 11-12, pp. 3113–3126, 2008. · ·

20.   R. Boukhanouf, A. Haddad, M. T. North, and C. Buffone, "Experimental investigation of a flat plate heat pipe performance using IR thermal imaging camera," Applied Thermal Engineering, vol. 26, no. 17-18, pp. 2148–2156, 2006. · ·

21.   Y. S. Muzychka, M. R. Sridhar, M. M. Yovanovich, and V. W. Antonetti, "Thermal spreading resistance in multilayered contacts: applications in thermal contact resistance," Journal of Thermophysics and Heat Transfer, vol. 13, no. 4, pp. 489–494, 1999.

22.   S. J. Kline and F. A. McClintock, "Describing uncertainties in single sample experiments," Mechanical Engineering, vol. 75, pp. 3–8, 1953.

Citations

CHAPTER 1

K. Ekambara, R. Sean Sanders, K. Nandakumar, and J. H. Masliyah, "CFD Modeling of Gas-Liquid Bubbly Flow in Horizontal Pipes: Influence of Bubble Coalescence and Breakup," International Journal of Chemical Engineering, vol. 2012, Article ID 620463, 20 pages, 2012. doi:10.1155/2012/620463.

CHAPTER 2

Bryan Poulson, "Predicting and Preventing Flow Accelerated Corrosion in Nuclear Power Plant," International Journal of Nuclear Energy, vol. 2014, Article ID 423295, 23 pages, 2014. doi:10.1155/2014/423295.

284

CHAPTER 3

Agustín Jose Torroba, Ole Koeser, Loic Calba, Laura Maestro, Efrain Carreño-Morelli, Mehdi Rahimian, Srdjan Milenkovic, Ilchat Sabirov, and Javier LLorca, Investment casting of nozzle guide vanes from nickel-based superalloys: part I – thermal calibration and porosity prediction, doi:10.1186/s40192-014-0025-5.

CHAPTER 4

L. Cantelli, A. Fichera, and A. Pagano, "A High-Resolution Resistive Probe for Nonlinear Analysis of Two-Phase Flows," Journal of Thermodynamics, vol. 2011, Article ID 491350, 10 pages, 2011. doi:10.1155/2011/491350.

CHAPTER 5

Chu-Hsuan Lee, Meng-Han Tsai, and Shih-Chung Kang, A Visual Tool for Accessibility Study of Pipeline Maintenance during Design, doi:10.1186/s40327-014-0006-y.

CHAPTER 6

M. R. Safaei, O. Mahian, F. Garoosi, et al., "Investigation of Micro- and Nanosized Particle Erosion in a 90° Pipe Bend Using a Two-Phase Discrete Phase Model," The Scientific World Journal, vol. 2014, Article ID 740578, 12 pages, 2014. doi:10.1155/2014/740578.

CHAPTER 7

Ali A. Abdul-Hadi and Anees A. Khadom, "Studying the Effect of Some Surfactants on Drag Reduction of Crude Oil Flow," Chinese Journal of Engineering, vol. 2013, Article ID 321908, 6 pages, 2013. doi:10.1155/2013/321908.

CHAPTER 8

R. Boukhanouf and A. Haddad, "Simulation and Experimental Investigation of Thermal Performance of a Miniature Flat Plate Heat Pipe," Advances in Mechanical Engineering, vol. 2013, Article ID 474935, 8 pages, 2013. doi:10.1155/2013/474935.

Index